Symmetry:
A Multi-Disciplinary
Perspective

Ramanujan Mathematical Society Lectures Notes Series

Ramanujan Mathematical Society

Lecture Notes Series

Volume 16

Symmetry: A Multi-Disciplinary Perspective

Volume editor

Inder Bir S. Passi
Indian Institute of Science Education and Research

International Press
www.intlpress.com

Ramanujan Mathematical Society
Lecture Notes Series, Volume 16
Symmetry: A Multi-Disciplinary Perspective

Volume editor:
Inder Bir S. Passi (Indian Institute of Science Education and Research, Mohali)

Published in 2012 by International Press, Somerville, Massachusetts, U.S.A. under license from the Ramanujan Mathematical Society.

This work is also published in the Republic of India, exclusively by the Ramanujan Mathematical Society.

ISBN: 978-1-57146-247-3

Printed in the United States of America.

16 15 14 13 12   1 2 3 4 5 6 7 8 9

Preface

The notion of symmetry is fundamental in all sciences, and indeed in all human intellectual endeavors. Its essence is captured through an abstract mathematical treatment based on the Theory of Groups. A Symposium with focus on the notion of symmetry as developed from a mathematical perspective and its interconnections with various other disciplines was held at the transit campus of the Institute of Science Education and Research Mohali located at MGSIPA Complex Sector 26 Chandigarh 160019 during 19–20 February 2010.

The Symposium was motivated by the fact that MTH 101: Symmetry is a mandatory course in the first year for all students of BS-MS at IISERM. While the faculty was fully convinced, the students often wondered about the need and justification for such a course in the very first year of their study. In order to enable the students to see the ubiquity of symmetry across various disciplines, experts from diverse fields were invited to the Symposium which was attended by the IISERM faculty and students.

With a view to make the deliberations at the Symposium available to a wider audience, it was decided to publish the proceedings as a volume which can serve as a supplementary reading for science students at undergraduate and post-graduate levels. It is expected that the articles by the leading experts in their respective fields will serve as a basis for designing instructional material aimed at exhibiting the unity of science. Perceptive readers will find numerous ideas which can be pursued as research projects.

The Organizing Committee is grateful to all the speakers and participants for accepting the invitation to participate, and for contributing their respective articles to this publication.

The Editorial Committee is thankful to the Ramanujan Mathematical Society, and, in particular, the Editor-in-Chief, Professor R. S. Kulkarni, for publishing the proceedings in the Lecture Notes Series in Mathematics.

Inder Bir S. Passi
IISER Mohali

Members of the Advisory Board

Foreword

Symmetry and Symmetry Breaking

When IISER Mohali was set up, there were no departments, schools or centres. The faculty were appointed in various disciplines. Sometimes jointly between two disciplines. Therefore, it was but natural that Professor Passi suggested that we have the first symposium in the Institute and that it be on symmetry. All of us agreed readily, as symmetry pervades all disciplines, living and nonliving, including mathematics and arts and architecture.

As a chemist, I see symmetry everywhere. Chemists classify molecules by symmetry groups. They classify solids by symmetry. More recently, they have started classifying atomic and molecular clusters by symmetry. They use point groups, translational groups and permutation groups as well. They work out selection rules based on symmetry as well as asymmetry. They are obsessed with symmetry as well asymmetry. Asymmetry does not mean not symmetric. It refers to two molecules which are mirror images but are not super imposable. Like the two hands of an individual. Chemistry is dynamic. Naturally, it deals with dynamical systems in which static symmetries are not adequate. One has to deal with dynamical symmetries.

Lots of things around us exhibit symmetry. But on close examination, one realises that the real beauty lies in breaking the symmetry. While the left side of the body looks like a mirror image of the right side, one realises that it is not one hundred per cent true. A chicken can be modelled as a sphere in the first approximation. But to represent a chicken, you need to destroy the spherical symmetry and build things around it!

Life is all about symmetry breaking. When one examines how life must have emerged, it becomes clear that asymmetric molecules must have played a crucial role. Sugars and all the naturally occurring amino acids except glycine are chiral (optically active, that is). Interestingly, the amino acids are of L type (levorotatory). The proteins that are formed out of these L-amino acids are helical and optically active.

How does asymmetry arise from symmetry? When molecules come together and form a crystal, they assemble in a particular way. Quartz is a classic example. Tartaric acid is another. Louis Pasteur could pick up crystals of different optical isomers of tartaric acid using tweezers.

It is known that hexagons can be combined to form a sheet as it happens in a graphite layer. But if you want to produce a football like structure as in fullerene, one needs to destroy the hexagonal symmetry and introduce a pentagon. Around a pentagon, hexagons can be put together to form a bowl. Once you form a bowl, you can add more pentagons and hexagons and produce a ball. Interestingly, that ball is not completely spherical. When Buckminster Fuller proposed the dome structure, nobody would have thought that the chemists would make molecules of that shape one day and start discussions on modifying their properties!

In this little note that I have prepared, I realise that I have touched upon physics, chemistry, biology and architecture, all in one go, by using symmetry and symmetry breaking.

I congratulate Professor Passi and his editorial team in putting together this volume on symmetry, the subject that cuts across disciplines and is vital for our understanding the world around us. I am grateful to all the authors for their valuable contributions. I am particularly happy that it has come from IISER Mohali showing the way for more volumes to come in the near future.

N. Sathyamurthy
Director
IISER Mohali

Invited Speakers

Professor R. S. Kulkarni
Visiting Professor
Department of Mathematics
Indian Institute of Technology Bombay
Powai, Mumbai 400 076.
email: kulkarni@math.iitb.ac.in

Professor S. Mukhi
Department of Theoretical Physics
Tata Institute of Fundamental Research
Homi Bhabha Road
Mumbai 400 005.
email: sunil.mukhi@tifr.res.in

Professor N. Mukunda
103, 6th Main Road Malleswaram
Bangalore 560 003.
email: nmukunda@ias.ernet.in

Professor M. Ramaswamy
Tata Institute of Fundamental Research
Centre for Applicable Mathematics
Sharada Nagar
Yelahanka New Town
Bangalore 560 065.
email: mythily@math.tifrbng.res.in

Professor V. Nanjundiah
Developmental Biology and Genetics
Laboratory, Indian Institute of Science
Bangalore 560 012.
email: vidya@ces.iisc.ernet.in

Ar Y. Pandya
Milan Bunglow, Sargam Flat Lane
Iswarbhuvan Stadium Road
Navjivan Post
Ahmedabad 380 014.
email: pandyatin@hotmail.com

Professor B. Sury
Indian Statistical Institute
8th Mile, Mysore Road
RV College Post
Jnana Bharathi, Bengaluru
Karnataka 560 059.
email: surybang@gmail.com

Professor K. Srihari
Department of Chemistry
Indian Institute of Technology
Kanpur 208 016.
email: srihari@iitk.ac.in

Participants

Faculty and students at IISER Mohali together with the guest speakers participated in the symposium. A special session for discussion and interaction with the students was organized.

Contents

Symmetry: A Multi-Disciplinary Perspective No. 16, 2011, pp. 1–12.

Development of the Idea of Symmetry: Examples from Geometry and Physics

Ravi S. Kulkarni

Department of Mathematics, Indian Institute of Technology (Bombay), Powai, Mumbai 400 076, India
e-mail: kulkarni@math.iitb.ac.in; punekulk@yahoo.com

Abstract. Space, Number, and Symmetry (SNS) are three categories of mathematical thought. We review how SNS are formulated in set-theoretic terms in mathematics. This wider formulation of "symmetry" brings mathematical precision to the study of the usual perception of symmetry of physical forms in Euclidean 3-space, based mainly on sense-perception.

We explain how this formulation developed historically through Euclidean, hyperbolic, and elliptic 2-dimensional geometries. We point out the implicit role which the understanding of real numbers played in this formulation. In Euclid's formulation, this role was implicit in his fourth postulate: "All right angles are congruent". We develop an example of a 2-dimensional Euclid's-like geometry based on a field, different from the field of real numbers, in which the Euclid's fifth postulate is true, but the fourth postulate is false. Real numbers are also important in getting the "uniqueness" assertion about Euclidean geometry, and other geometries. In the last section, we briefly explain Lie groups, and the classification of simple, and low-dimensional, Lie groups, and point out their role in modern physics and differential geometry, in particular the uniqueness assertion about Euclidean and other geometries.

Keywords. Categories of mathematical thought; Euclidean, hyperbolic, and elliptic 2-dimensional geometries; Euclid's fourth and fifth postulates; anisotropic quadratic form; Lie groups.

1. Introduction

Space, Number and Symmetry are three categories of mathematical thought. That is how we, often subconsciously, think in mathematics before we find expression for our thought using a widely agreed upon language of mathematics. Until the mid-nineteenth century, only space and number were recognized as categories of mathematical thought. But after the works of Galois, Jordan, Frobenius, Burnside and others, on one hand, and the works of Lie, Killing, Cartan, and others, on the other, Hermann Weyl cogently proposed that symmetry should also be recognized as a category of mathematical thought. The proposal is now well accepted, as is clear from the title of this symposium.

The Space-Number-Symmetry (SNS) categories of mathematical thought may be compared to the categories of physical thought, namely Space, Time and Matter (STM). But even STM are finally interpreted in terms of SNS. As we have come to recognize, a re-interpretation of STM in terms of SNS in the twentieth century, arising with association of the concept of time with light-signals, and the experimental fact that the speed of light is finite and appears to be constant, has led to a revolution in physics.

The "space" in mathematics interpreted in terms of sets and structures on sets has a wider connotation than the "space" in physics, which is primarily the 3-dimensional space of our physical experience, and the spaces arising therefrom. The notions of sets with structures, and the morphisms of such objects, are a much wider and clearer mathematical formulation of what is described as the "statics and dynamics of particles" in physics. The notions of physical reality, based mainly on sense-perceptions, are replaced by existence of appropriate sets and allowable constructions on them. Apart from this, space, number, time, matter have been present in all theory-building in mathematics, physics, chemistry and biology, quite consciously. But "symmetry" is also present, often subconsciously, in all theory-building. In mathematics, "symmetry" is formulated very precisely in the language of groups and their actions on sets. This formulation goes well beyond just the symmetries of physical forms in the 3-dimensional Euclidean space. It would be interesting to investigate the number of Fields Medalists and Nobel Laureates, and other distinguished mathematicians, physicists, chemists, and biologists in the 20th century, in whose works "symmetry" was used, often subconsciously, in a crucial way. See section 4, for the role of Lie groups in modern mathematics and physics.

In this lecture I shall attempt to illustrate this aspect in the discussion of the history of some classical geometries and their use in physical theories, from a modern perspective.

The controversy regarding the independence of the 5th postulate from the other four postulates in the classical plane Euclidean geometry is well-known. I shall look at this controversy from the perspective of symmetry. In the discussion of this controversy, the underlying role of the 4th postulate in not yet fully recognised. In this largely expository paper, we shall point out that the 4th postulate has deep connections with "what are or what should be the real numbers" – to use the title of a famous paper of Dedekind. If we allow ourselves to build Euclidean geometry based on other number systems, for example, rational numbers system which has richer arithmetic, then we can get "generalised Euclidean geometries" where the 5th postulate holds, but the 4th postulate fails to hold. Moreover, there may arise additional "arithmetic" invariants, which may lead to infinitely many geometries. The only arithmetic of "real" numbers à la Dedekind, is whether the number is positive or negative. So the 4th and the 5th postulates, based on Dedekind's real numbers, suffice to ensure the existence as well as uniqueness of the plane Euclidean geometry. This point has been elaborated in an unpublished manuscript by the author and Larry Mansfield.

2. Euclidean, hyperbolic and elliptic 2-dimensional geometries

At the outset, I should mention that Euclid's formulation of his geometry is what we would now call Euclidean similarity geometry, rather than Euclidean metric geometry. For the notion of metric, i.e. distance, we need real numbers. The "reality" of irrational numbers (i.e. "numbers" which cannot be expressed as ratios of two natural numbers) baffled the Greeks of Euclid's days, as indeed it baffled the mathematicians up to the 19th century. There was also no clarity about the issue of positive and negative numbers. Euclid would rather formulate positive real numbers, as "similarity factors of similarity transformations". Also the Greek notation for expressing numbers lacked the finesse of the decimal system.

All notions introduced by Euclid are spatial in character. For example, for Euclid, "angle" is *not* a number. It is the space between two rays.

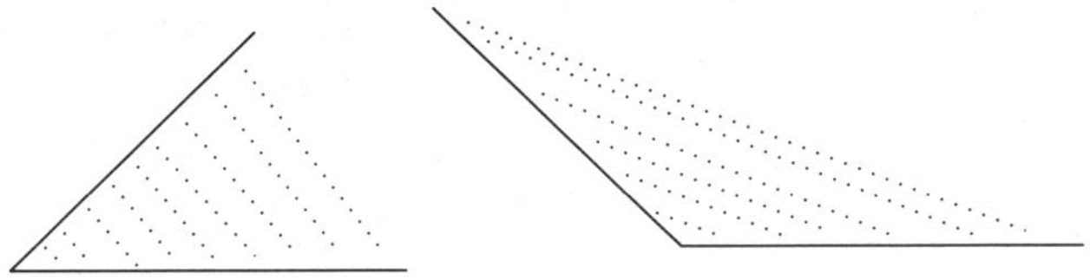

"Angle" is a Latin word. The Greek equivalent is "gonon", which is a cognate of the Sanskrit word कोण. Thus, came the words like triangle, quadrangle, pentagon, hexagon, etc., or त्रिकोण, चतुष्कोण, पंचकोण, षट् कोण, etc. The word "tri-gono-metry" (त्रि-कोण-मिति) refers to rules assigning measures i.e. numbers, to three corners of a triangle.

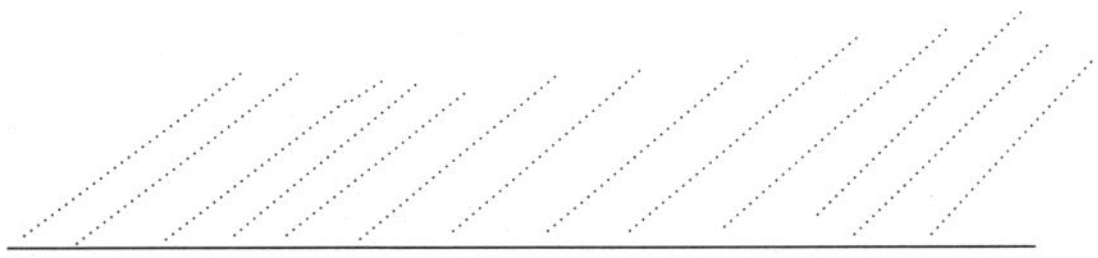

A "straight angle" for Euclid is *not* 180°, or π radians. It is the region formed by two rays when they form a line.

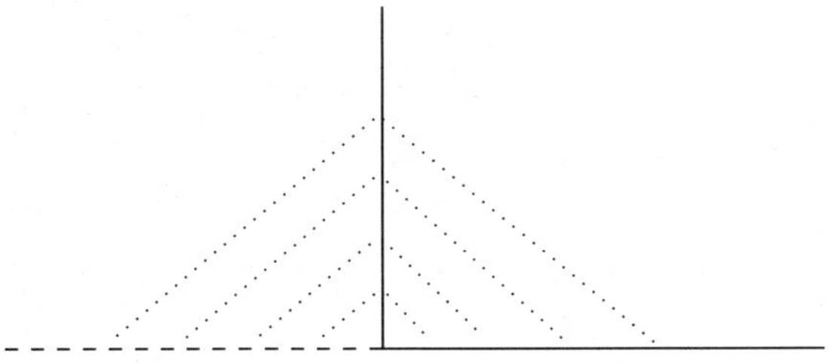

A "right angle" is *not* 90°, or $\frac{\pi}{2}$ radians. It is the region formed by two rays, which is *congruent* to the complementary region, formed by one of the rays when the other ray is extended to form a line. If a carpenter wishes to build a wooden wall over the ground, he would place it at a "right" (i.e. correct) angle to the ground. *Congruence* is an undefined term in Euclid, and contains the germ of the idea of "distance"-preserving bijections, which form a group in modern terminology.

Euclid built his geometry based on the following postulates. Let us try to formulate them in set-theoretic terms. Let X be a set, which we call a "plane". The set $\mathcal{L}$ of certain subsets of X is called the set of "lines". The set $\mathcal{C}$ of certain subsets of X is called the set of "circles". It is assumed:

1) For every $P, Q \in X$, $P \neq Q$, there exists a unique $l \in \mathcal{L}$ such that $P \in l$, and $Q \in l$.

Before coming to further postulates let us note that as Hilbert pointed out in his "Foundations of Geometry", in purely set-theoretic terms, we need to add much more to Euclid's original expression of his postulates. Euclid has before his mind's eye, a sheet of paper, which he assumes, is extending indefinitely "beyond horizon".

2) Postulate of infinite extent: Every $l \in \mathcal{L}$ extends indefinitely in both directions.
3) A *radius* is a finitely extended line-segment. For every $O \in X$ and radius R, there exists a *unique* $C \in \mathcal{C}$ with "centre" O and radius R; and for every $C \in \mathcal{C}$ there exists a unique "centre" $O \in X$ and a unique radius R.

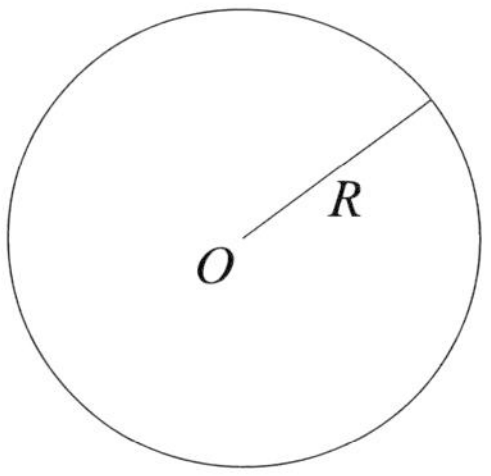

4) All right angles are *congruent*.
5) Parallel Postulate: for every $l \in \mathcal{L}$, and $P \notin l$, there exists a *unique* $m \in \mathcal{L}$ such that $P \in m$, and l is *parallel* to m (meaning $l \cap m = \phi$).

Later on Euclid also uses the word *similar*, as in "similar triangles". Implicitly, he is using two equivalence relations on the set of subsets of X, which, from a modern perspective, are best formulated in terms of the *"isometry"* group and *"similarity"* group, and their *actions* on X. There is also a very subtle issue of *orientability*. Euclid talks of "similar" triangles, but nowhere does he mention "similarly oriented" triangles. This issue got satisfactorily resolved only after the advent of algebraic topology.

It seems that Euclid wishes to avoid the use of the parallel postulate as much as possible. He crucially uses it in the proof

($*$) *The three angles (corners) of a triangle can be assembled to form a straight line.*

The proofs are indicated in the figures below:

There was a lot of activity extending over millenia upto the beginning of the nineteenth century to deduce 5) from 1) – 4). The great philosopher Kant even asserted that 5) is an *a priori knowledge*, so it is "true". However, Lobachevsky and independently Bolyai, around 1830*s* constructed geometries in which 1) – 4) are valid but 5) is false. From a modern perspective they discovered a new geometry, which we now call *hyperbolic geometry*. In hyperbolic geometry, the angle-sum of a triangle is less than π. For emphasis, note again that the words "angle-sum" or "less than π" do not

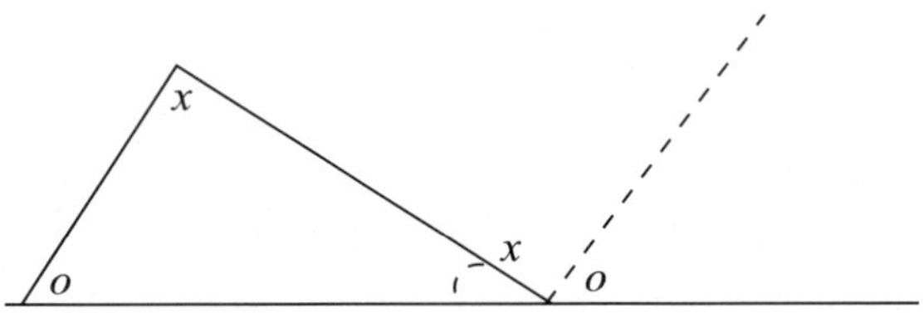

OR

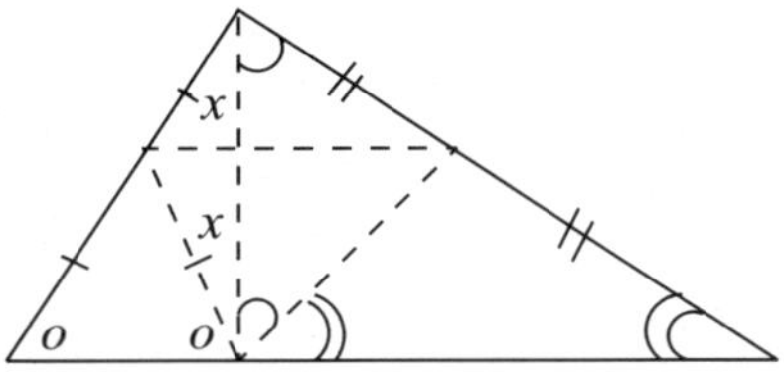

refer to addition of numbers or any specific number. The "angle-sum" refers to the corners being assembled around a vertex, and "less than π" refers to the statement that this assemblage of "angles" occurs within a straight angle.[1]

Poincaré model for 2-dimensional hyperbolic geometry

Let X be an open round disk in the Euclidean plane $\mathbb{E}^2$. Let $\bar{X}$ denote its closure. $\partial X = \bar{X} \backslash X$ is a round circle. X is eventually equipped with a hyperbolic metric, in which ∂X "exists at infinity", but has no meaning in terms of distances in X. But ∂X has a *conformal* meaning. Let O be the Euclidean center of X. It will turn out that O also cannot be given any special status in hyperbolic geometry.

For every $P, Q \in X$, $P \neq Q$, s.t. P, Q, O are non-collinear, there exists a unique Euclidean circle in $\mathbb{E}^2$ which cuts ∂X at "right" angles. If P, Q, O are collinear, then the Euclidean line formed by them cuts ∂X at "right" angles. Let the open arc of the circle through P and Q lying in X in case P, Q, O are non-collinear, and the open segment of the line through P, Q in case P, Q, O are collinear, be called *the (hyperbolic) line through P and Q*. Thus we get the set $\mathcal{L}$ of hyperbolic lines.

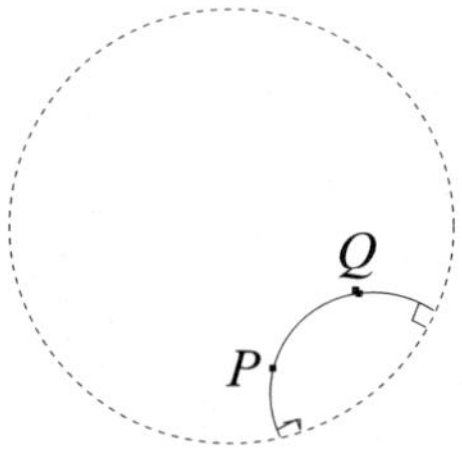

[1]Gauss had a rather ugly priority battle with Bolyai who was the son of his friend, and Lobachevsky, regarding the discovery of hyperbolic geometry. He surely understood the issues. After all the "Gauss-Bonnet" theorem, (which again was not stated explicitly by Gauss, but the Gauss' paper of 1827 contains the essential ideas) is a penetrating and far-reaching extension of the angle-sum theorem. Still, the secrecy in scientific research, its communication, and the priority battles, is part of academic politics, which has continued to our times. Perhaps the twenty-first century will witness a more mature view of the role of priority in scientific research, and the value of its rapid dissemination, for the benefit of the society at large.

[*Construction of the circle through P, Q when P, Q, O are non-collinear*: Given $P \in X$, $P \neq O$, its inverse P^* with respect to ∂X in $\mathbb{E}^2$ is constructed as follows: In case P, Q, O are non-collinear, the circle through P, Q which cuts ∂X at right angles is the circle through P, P^*, Q, Q^*. (It is easily proved that these four points do lie on the circle). In case P, Q, O are collinear, P, P^*, Q, Q^* lie on an Euclidean line which cuts ∂X at right angles.

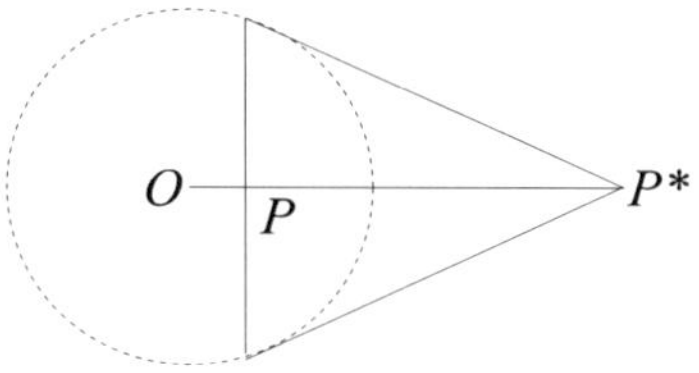

The inverse of O is left un-defined by this procedure. In *inversive or Möbius geometry* the inverse of O is taken to be "the point at infinity", which is a new point adjoined to $\mathbb{E}^2$. The inverse of the point at infinity is defined to be O. The function $P \rightarrow P^*$ is the inversion in ∂X. The inversion can, of course, be defined w.r.t. any Euclidean circle. An inversion in an Euclidean line is just the Euclidean reflection in the line. The important fact about inversions is that they fix the circle (line) in which we are inverting, carry circles and lines in $\mathbb{E}^2$ into circles and lines in $\mathbb{E}^2$, and preserve angles. In particular, an inversion in a circle C leaves invariant every circle which cuts C at "right" angles.]

The set $\mathcal{C}$ of circles in X is just the set of Euclidean circles in $\mathbb{E}^2$ contained in X. However, if C is a circle, its Euclidean center and "radius" are different from its hyperbolic center and radius. We skip these delicate points here, but note the important fact that the group generated by inversions in hyperbolic lines (which are mostly circles in $\mathbb{E}^2$) leaves X invariant. The invariance of X with respect to an inversion in a hyperbolic line which is also an Euclidean line is obvious. The inversion in a hyperbolic line which is the arc of an Euclidean circle, also leaves X invariant, because by definition, a hyperbolic line cuts X at "right" angles."Congruence" in hyperbolic geometry is defined with respect to this group generated by inversions in hyperbolic lines.

We easily see that postulates 1) – 4) are satisfied. But postulate 5) does not hold. Indeed

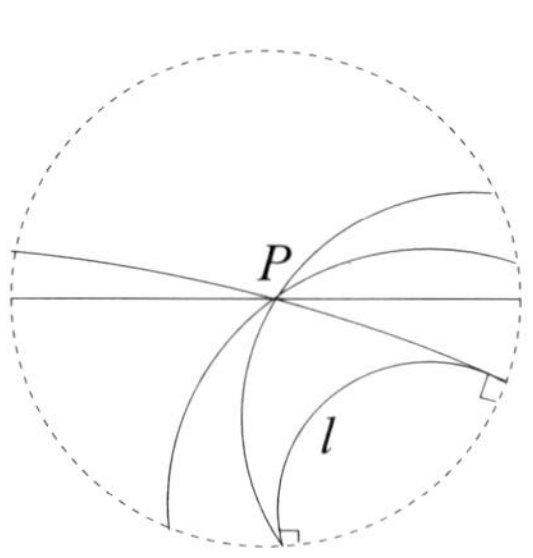

Given $l \in \mathcal{L}$, and $P \notin l$, there are infinitely many $m \in \mathcal{L}$ such that $P \in m$ but $l \cap m = \phi$. In fact there are two special lines through P which do not intersect l. For each end point of l (in ∂X), there is a unique line through P that is tangential to l at this end-point. This gives two lines through P which can be called the *parallel* lines through P to l in the non-Euclidean sense.

Elliptic geometry

Let S^2 be the unit sphere in $\mathbb{E}^3$. A "great circle" is the intersections with S^2 a plane in $\mathbb{E}^3$ through the centre of S^2. Tentatively, call a great circle in S^2, a *line* in S^2. All other intersections of planes at distance strictly less than 1 and strictly greater than 0 from the centre are called *circles* in S^2. The "geometry" thus constructed does not satisfy 1). For, a pair of antipodal points are joined by infinitely many lines.

There is a simple trick to remove this defect. Consider the space

$$\mathbb{P}^2 = S^2/\sim$$

where $\sim$ identifies antipodal points. That is, $x \in S^2 \sim -x \in S^2$. A pair of antipodal points is counted as one point in $\mathbb{P}^2$. The projection of a line (respectively circle) in S^2 is called a line (respectively circle) in $\mathbb{P}^2$. Now 1) holds! (Notice the power of basing mathematics on set theory, and assigning "reality" to set-theoretic constructions. It is completely irrelevant, whether we can "see" $\mathbb{P}^2$ in $\mathbb{R}^3$ at all. In fact it can be proved that we shall never be able to see $\mathbb{P}^2$ in $\mathbb{R}^3$.)

Let

$$G = \{A : \mathbb{R}^3 \longrightarrow \mathbb{R}^3 | A \text{ a linear transformation}, \ A^t A = I\}$$

G is isomorphic to the group $O(3)$ of all 3×3 orthogonal matrices. G acts on $\mathbb{R}^3$ and preserves S^2. Moreover, G preserves anti-podality, i.e. pairs of antipodal points go to pairs of antipodal points. So G acts on $\mathbb{P}^2$. "Congruence" is defined with respect to the G-action.

The hypothesis 4)–namely "all right angles are congruent" amounts to: G acts transitively on lines. So 1) – 4) essentially hold. But 5) does not. In this geometry any two lines intersect! This is also a non-Euclidean geometry.

3. The role of the 4th postulate

Recall the 4th Postulate

4) : All right angles are congruent.

Actually, as was mentioned earlier, Euclid uses two equivalence relations, namely congruence and similarity on subsets of $\mathbb{E}^3$. But in order to avoid the use of real numbers, he does not formulate these notions in terms of distance.

It may be remarked, that Euclid's proof of the Pythagoras theorem is *not* an equality of numbers "$c^2 = a^2 + b^2$". He argues that the square built on the hypotenuse of a right-angled triangle can be dissected and re-assembled to form squares on the remaining sides a, b. Bhaskaracharya's proof uses the same idea, and is simpler (see below). The quadratic nature of the Euclidean distance was recognized as a specific hypothesis on the "space" much later (after Gauss' work on curved surfaces) by Riemann (1851).

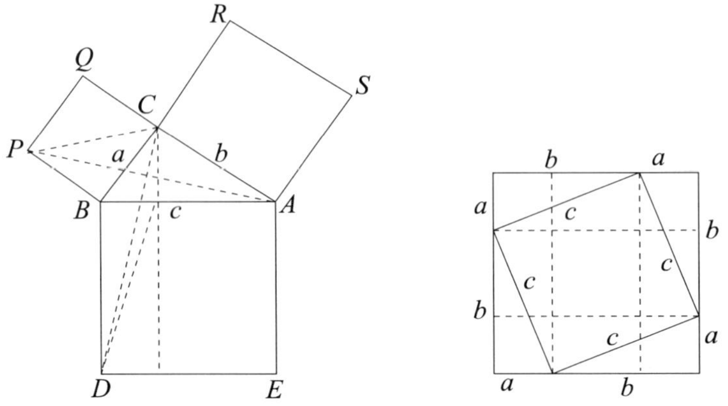

Euclid's proof Bhaskaracharya's proof

So far "congruence" is an undefined term. Euclid seems to have thought of "congruence" in physical terms: if a subset A can be moved "rigidly" to a subset A' – on the sheet of paper before him – then A is congruent to A'. With S for side and A for angle, SSS and SAS are the usual tests mentioned in high school geometry for congruence of two triangles.

While AAA is a test for similarity of two triangles.

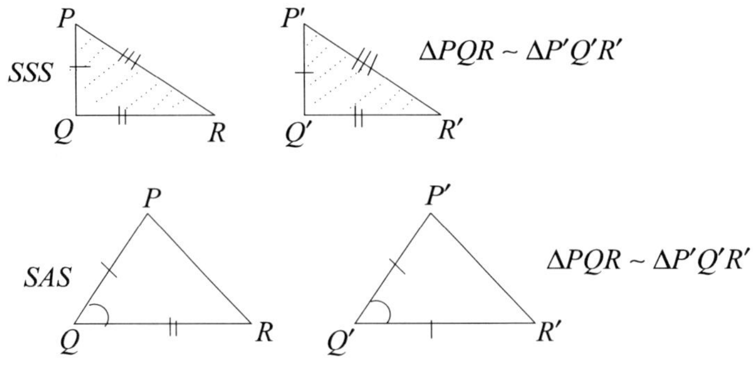

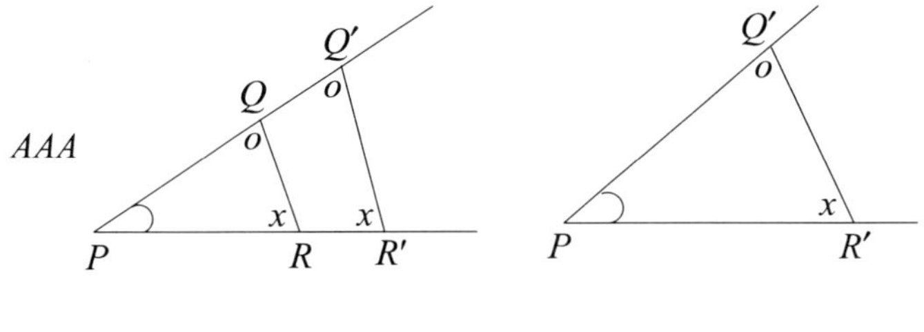

Euclid's 4th postulate is his expression for his perception of underlying symmetry. After two millenia, the mathematicians have found a better expression in the language of groups and their actions, which applies to many other situations. For this to happen, it was fortuitous that the issue of what "are" or what "should be" the real numbers was resolved independently of similiarity geometry–as was eventually done in the work of Dedekind (1880). Hilbert's "Foundations of Geometry" was based on Dedekind's work.

Taking $\mathbb{R}$ as a starting point, and fixing a point $O \in \mathbb{E}^2$, we can turn $\mathbb{E}^2$ into a vector space $\mathbb{V}$ over $\mathbb{R}$. We can equip $\mathbb{V}$ with a positive definite quadratic form

$$\beta : \mathbb{V} \times \mathbb{V} \longrightarrow \mathbb{R}$$

Let $\mathbb{A}$ be the underlying *affine* space to $\mathbb{V}$, and define distance by

$$d(\vec{v}, \vec{w}) = \sqrt{\beta(\vec{v} - \vec{w}, \vec{v} - \vec{w})}$$

Or, in set-theoretic terms, we can take $(\mathbb{A}, d)$ as the *definition* of $\mathbb{E}^2$. [A subtle point: the expression $\vec{v} - \vec{w}$ does not depend on the choice of the point O, used to define the vector space.]

Definition.

(i) *A bijection $f : \mathbb{E}^2 \longrightarrow \mathbb{E}^2$ is an isometry if for every $P, Q \in \mathbb{E}^2$,*

$$d(f(P), f(Q)) = d(P, Q)$$

(ii) *f is called a similarity if there exists $r \in \mathbb{R}_{>0}$, called the "similarity factor" of f, such that for every $P, Q \in \mathbb{E}^2$,*

$$d(f(P), f(Q)) = rd(P, Q).$$

Define the groups,

$$\mathcal{E}(2) = \{f : \mathbb{E}^2 \longrightarrow \mathbb{E}^2 | f \text{ is an isometry}\}$$

and

$$\mathcal{S}(2) = \{f : \mathbb{E}^2 \longrightarrow \mathbb{E}^2 | f \text{ is a similarity}\}$$

These sets become groups under composition. And finally,

Definition. *Two subsets $S_1, S_2 \subseteq \mathbb{E}^2$ are congruent (resp. similar) if there exists $f \in \mathcal{E}(2)$ (resp. $\mathcal{S}(2)$) such that $f(S_1) = S_2$.*

With these definitions, Euclid's postulate 4) becomes:

Proposition. *Given two right angles A_1, A_2, there exists $f \in \mathcal{E}(2)$ such that $f(A_1) = A_2$*

Remarks.

(i) In Minkowski's (1908) formulation of special relativity, the analogue of Euclid's postulate 4) is precisely: the speed of light (for two observers in uniform motion with respect to each other) is constant. With more input from physics, it is known that in $3 + 1$ dimensions of special relativity, the analogue of Euclid's postulate 4) is equivalent to Maxwell's equations. An otherwise complicated theory, which explains enormous chuck of physical experience, is so beautifully and succinctly explained once one becomes conscious of the underlying symmetry!

(ii) In our description of the hyperbolic and elliptic geometries we again invoked certain groups to define congruence. The word "congruence" in postulate 4) is interpreted in group-theoretic terms. However, there are no non-isometric similarities in these geometries.

(iii) Let $\mathbb{F}$ be a field of characteristic $\neq 2$. Let $\mathbb{V}$ be a 2-dimensional vector space over $\mathbb{F}$. Suppose there exists a quadratic form

$$\beta : \mathbb{V} \times \mathbb{V} \longrightarrow \mathbb{F}$$

such that for every $\vec{v} \in \mathbb{V}$, $\vec{v} \neq \vec{0}$ we have $\beta(\vec{v}, \vec{v}) \neq 0$. Consider the underlying affine space $\mathbb{A}$. Define $d : \mathbb{A} \times \mathbb{A} \longrightarrow \mathbb{F}$ as

$$d(\vec{v}, \vec{w}) = \beta(\vec{v} - \vec{w}, \vec{v} - \vec{w})$$

[We do not take $d(\vec{v}, \vec{w}) = \sqrt{\beta(\vec{v} - \vec{w}, \vec{v} - \vec{w})}$ for the square-roots may not exist in $\mathbb{F}$.] Define a *generalized Euclidean metric geometry over* $\mathbb{F}$ *with respect to* β : $(\mathbb{A}, d) =: \mathbb{E}^2_{\mathbb{F},\beta}$. And,

$$\mathcal{E}(\mathbb{E}^2_{\mathbb{F},\beta}) := \{f : \mathbb{E}^2_{\mathbb{F},\beta} \longrightarrow \mathbb{E}^2_{\mathbb{F},\beta} \mid f \text{ preserves } \beta\}$$

Definition. $\vec{v}, \vec{w} \in \mathbb{V}$ *form a right-angle if* $\vec{v}, \vec{w}$ *are linearly independent, and* $\beta(\vec{v}, \vec{w}) = 0$.

In this geometry the *postulate 4) fails in general*. In fact, let Δ be the discriminant of β. Consider the field extension $\mathbb{E} = \mathbb{F}(\sqrt{-\Delta})$. Then the orbits of $\mathcal{E}(\mathbb{E}^2_{\mathbb{F},\beta})$ on the set of right angles form a group, which is isomorphic to $N(\mathbb{E}/\mathbb{F})/\mathbb{F}^{*2}$. Where

$$N(\mathbb{E}/\mathbb{F}) = \text{norms on } \mathbb{E}/\mathbb{F} = \{x^2 + \Delta y^2 | x, y \in \mathbb{F}, (x, y) \neq (0, 0)\}$$

is a group. And $\{x^2 | x \in \mathbb{F}, x \neq 0\}$ is a (possibly proper) subgroup of $N(\mathbb{E}/\mathbb{F})$.

For $\mathbb{F} = \mathbb{R}$ and $\beta =$ the usual inner product, fortunately this does not occur. For in this case, the discriminant of β, Δ is 1. So $\mathbb{E} = \mathbb{R}(\sqrt{-1})$ is isomorphic to $\mathbb{C}$, the complex numbers, and

$$N(\mathbb{C}/\mathbb{R}) = \{x^2 + y^2 | x, y \in \mathbb{R}, (x, y) \neq (0, 0)\} = \{x^2 | x \in \mathbb{R}, x \neq 0\}$$

For $\mathbb{R}$ has the fortuitous property that $r \in \mathbb{R}$ is a square if and only if $r \geq 0$.

Such remarks go back to Dedekind. They occur in the introduction of his famous paper "What are, or what should be, the real numbers" (1880). But the role of postulate 4) is probably not yet sufficiently recognised.

To summarize and emphasize: given a field $\mathbb{F}$ of characteristic $\neq 2$, $\mathbb{V}$ a 2-dimensional vector space over $\mathbb{F}$ and $\beta : \mathbb{V} \times \mathbb{V} \longrightarrow \mathbb{F}$ an *anisotropic* quadratic form (i.e. for all $\vec{v} \in \mathbb{V}\backslash\{\vec{0}\}$, $\beta(\vec{v}, \vec{v}) \neq 0$). It is possible to construct *generalized Euclidean metric geometries* $\mathbb{E}^2_{\mathbb{F},\beta}$ *in which Euclid's 4th postulate does not hold.* However, it is possible to define a larger group $\mathcal{S}(2)$ with respect to which there is a *unique generalized similarity geometry,* $\mathbb{E}^2_{\mathbb{F},\beta,sim}$, *in which all right-angles are similar.*

These considerations also underline the role of real numbers in Euclidean thought, which was expressed better by Dedekind many years later. Intuitively assuming real numbers, Lobachevsky-Bolyai-Gauss constructed non-Euclidean geometries. However the issue of *uniqueness* of Euclidean, Hyperbolic and Elliptic geometries was not addressed until the role of Lie groups (based on real numbers) became clear in the late nineteenth century.

It should be remarked that in the algorithmic sense of Gödel-Turing, most real numbers are fictitious! In fact if you believe the physicists, the number of atoms, and elementary particles, in the universe is finite. So even the natural numbers $1, 2, 3, \ldots$ cannot be simulated by a material model. So do natural numbers exist? Are Dedekind's real numbers "real"? A famous statement of Kronecker (1895) is that the natural numbers were created by God, everything else is man's work! (If you believe that "God" and "Man" are not different, it follows that everything is man's or God's work!). A precise assertion by Dedekind on this issue is that "the real numbers are creations of the human mind".

But the precise *uniqueness* assertions of Dedekind and Cantor about the real numbers, and the underlying real line, should suffice to believe in their existence, i.e. their "reality". So perhaps Kant, in a literal sense, was wrong in asserting that postulate 5) and the Euclidean geometry were parts of a priori knowledge. But in a broader sense, he was right if we put "real numbers" and "real line" in a priori knowledge, on which the 2-dimensional Euclidean, hyperbolic and elliptic geometries, and their higher dimensional analogues, all explained in terms of Lie groups, are based.

We now turn to the precise uniqueness statements about these geometries.

4. A modern perspective: The role of Lie groups

As we have seen, and as was remarked by Prof. Mukunda in his lecture, that the mathematicians, and physicists, have come around to express the thought of symmetry, in the language of groups and their actions.

Among the groups, two classes stand out:

1) Finite groups.
2) Lie groups : examples $(\mathbb{R}, +)$, $\mathbb{R}^*$, $O(3)$, $GL_n(\mathbb{R}), \ldots$

It is not important, in my opinion, to know the definition of a Lie group. It is better to learn this subject through examples.

Recall that a group is called *simple* if it does not have proper normal subgroups. Among the major achievements of 20th century mathematics are the precise classification theorems of *finite simple groups* and *connected simple Lie groups*. Certain persons who made major contributions in these areas, e.g. Killing, Cartan, Weyl, Chevalley, Thompson, Ashbacher, ... stand out. But largely this was a grand team work. Not individual glory, but team-work, seems to be a growing and healthy trend in all scientific research, and its applications, in the 21st century.

We now mention a first few steps in these classification theorems, and how they helped summarize and suggest a vast amount of mathematics and physics. (Some of the examples from physics, in more detail, already occurred in Prof. Mukunda's lecture.)

Let G be a group, and H a subgroup. Then G acts on $X = G/H$. The action is transitive. So G/H are natural candidates for modelling geometries which exhibit a lot of symmetry. For example $\mathbb{E}^2$ can be realized as $\mathcal{E}(2)/O(2)$. We list a few low-dimensional examples below. From the classification of connected simple Lie groups we know:

(I) The lowest dimension of a simple Lie group is 3. The 3-dimensional connected, simple Lie groups are: $SO(3)$, $PSl_2(\mathbb{R}) \approx SO_0(2, 1)$. These are precisely the groups of 2-dimensional elliptic and hyperbolic geometry.

(II) The next dimension is 6. The 6-dimensional connected, simple Lie groups are closely related to: $SO(4)$, $SO(3, 1)$, $SO(2, 2)$.

  (i) $SO(4)$ is the group of 3-dimensional elliptic geometry.
  (ii) $SO(3, 1)$, whose identity component is isomorphic to $PSl_2(\mathbb{C})$, is simultaneously the group of 3-dimensional hyperbolic geometry, 2-dimensional inversive (Möbius) geometry and $3 + 1$ dimensional special relativity.
  (iii) $SO(2, 2)$ is the group of 3-dimensional de Sitter space, the model space for $2 + 1$ dimensional "space-time" of non-zero constant curvature.

(III) The next dimension is 8. A significant 8-dimensional simple Lie group is $PSU(3)$. It is the group of the standard model of quantum mechanics.

(IV) The connected Lie groups of dimension at most 4 are also classified. In particular, the only 3-dimensional Lie groups which contain $SO(2)$ as a subgroup, and have trivial centre, are $\mathcal{E}(2)$, $SO(3)$, and $PSl_2(R)$. The 2-dimensional Euclidean, elliptic, and hyperbolic geometries can be constructed by these groups alone. This constitutes the uniqueness assertion of these three geometries.

There is enough material here to keep oneself meaningfully busy, and gainfully employed!

Symmetry: A Multi-Disciplinary Perspective No. 16, 2011, pp. 13–24.

The Predictive Power of Symmetries: Lie Algebras, Super-Algebras and 3-Algebras in Physics

Sunil Mukhi

Tata Institute of Fundamental Research, Homi Bhabha Rd, Mumbai 400 005, India
e-mail: mukhi@tifr.res.in

Abstract. The postulate of local symmetry predicts new physical phenomena. This approach to formulating theories of nature has led to striking successes in understanding the electro-weak and strong interactions during the last three decades of the 20th century. Today, generalisations of the idea are poised for experimental confirmation and could provide a deeper understanding of how these fundamental forces are unified.

Keywords. Symmetry, gauge principle, elementary particles, supersymmetry.

1. Introduction

Symmetry in physics is normally incorporated through the mathematical concept of groups and algebras. Symmetry can be thought of as an "operation" that acts on objects (or systems), transforming them into some new objects. If the properties of these new objects remain the same as the old ones then we have a symmetry, or "invariance". The operation can be thought of as an abstract element of a set, and physical requirements make it clear that this set must have a multiplication (the composition of successive operations), an inverse (the undoing of an operation), and an identity (no operation). Moreover, three successive operations can be understood by combining the first two and then the third, or the last two and then the first. All these physically evident properties translate into the axioms of a group [1].

Once we consider continuous transformations that can be built up in small stages, the natural language becomes that of algebras. These arise in physics when we consider the way *infinitesimal* (arbitrarily small) transformations act on objects or systems. This article is about the role of Lie algebras and superalgebras in certain areas of physics, mainly particle physics and string theory [2].

The focus will be on *phenomena predicted by symmetry*. Indeed, the theme will be the inference:

$$\text{Continuous symmetry} \Rightarrow \text{Physical phenomena}$$

At first sight this seems quite surprising. One would think a physical system either has a symmetry or does not. That is of course true. But in particle physics we do not know everything about the system – the collection of all possible states of all fundamental particles – to start with. Experiments help us uncover only pieces of it at a time. To make a sensible theory from experiments, it helps to consider the possibility of various symmetries (that are in turn suggested by experimental facts).

These quite naturally constrain the possible theories we can formulate. But as the following discussion will show, these constraints can even be powerful enough to predict new particles, forces and physical effects.

In addition to Lie algebras, that naturally derive from continuous (and differentiable) groups, there are structures called superalgebras that arise by modifying some of the basic assumptions of Lie algebras. Their role in physics is not yet clear, but they are of great potential importance, with the imminent possibility of experimental discoveries that could establish their relevance in nature. A far more speculative extension of Lie algebras, known as 3-algebras, has attracted much attention in recent years and is the subject of my own work. This will be discussed towards the end.

This article can be seen as an elaboration of the third role of symmetry, the *creative* role, as discussed in Prof. Mukunda's beautiful article in this volume [3].

2. Translations and rotations

In nature we often encounter continuous transformations of a system. These can be performed infinitesimally (a little bit at a time). Translation is a simple example:

$$\delta\vec{x} = \vec{a} \tag{2.1}$$

where $\vec{a}$ labels a small constant vector by which the initial point $\vec{x}$ is translated.

In Euclidean space, combining translations in different orders leads to the same result. If T_1 and T_2 are two translations then this fact is represented by:

$$T_1 T_2 = T_2 T_1 \tag{2.2}$$

Bringing both terms to one side we have:

$$[T_1, T_2] \equiv T_1 T_2 - T_2 T_1 = 0 \tag{2.3}$$

where the square brackets are simply shorthand for the left hand side of the equation. The transformations are said to "commute" and the corresponding algebra is called Abelian.

Suppose we now perform rotations. We have a choice of three rotation planes and corresponding rotation generators:

$$\begin{array}{ccc} (x, y) & (y, z) & (z, x) \\ R_{xy} & R_{yz} & R_{zx} \end{array} \tag{2.4}$$

We illustrate this by rotating a picture of the physicist Vladimir Fock, whose ideas will play an important role in the following sub-section.

An important point must be noted here[1]. The rotated picture is not identical to the original one. It could only have been so if Dr. Fock's head was perfectly spherical! However, our theories of physics *are* invariant under such a rotation. This exemplifies the fact that invariance under symmetry transformations is a property of an underlying

[1] This is explained more thoroughly in Prof. Mukunda's article in this volume.

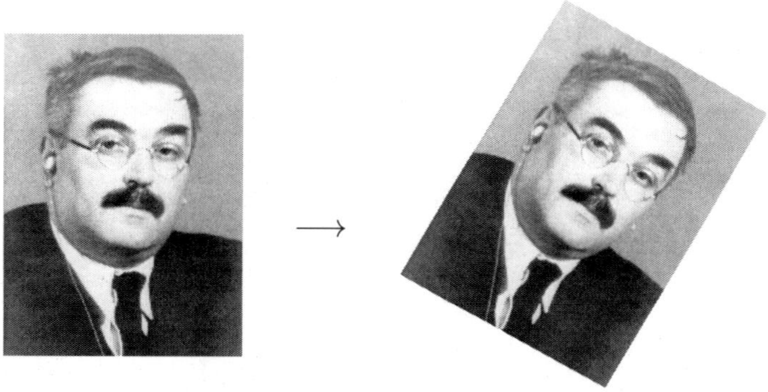

Figure 1. Example: rotating Dr. Fock.

theory but not usually of specific states in the theory. Instead of being invariant, states generally transform into each other. A simple context in which to appreciate this point is translation invariance: the laws of nature are invariant under translations of space, but physical objects are not translation-invariant! Instead, translations take them to different locations.

As we know from experience, the rotation generators do not commute. If we rotate successively in different planes, the order of rotation is important. In fact, if R_{ij} denote the generators of rotations in three different planes (where $i, j \in x, y, z$) then we have:

$$[R_{xy}, R_{yz}] = R_{xz} \tag{2.5}$$

and obvious cyclic permutations of this. Such algebras are called non-Abelian. This particular one is called $O(3)$.

3. Lorentz symmetry

3.1 *Global version*

With the discovery of special relativity in 1905, the rotation algebra was enhanced to a larger algebra containing both rotations and Lorentz boosts. The boosts take us from one inertial reference frame to another. In some ways (not all) they behave

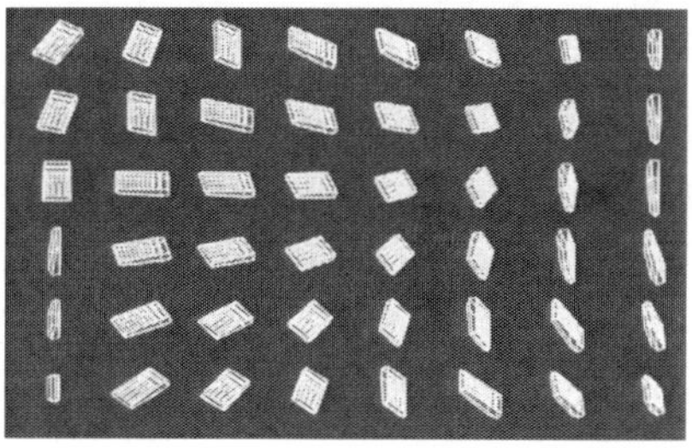

Figure 2. A visual impression of rotations[2].

like generalised "rotations" between time and space. Thus we introduce the boost generators:

$$R_{xt} \qquad R_{yt} \qquad R_{zt} \tag{3.1}$$

satisfying relations like:

$$[R_{xt}, R_{yz}] = 0, \quad [R_{xt}, R_{xy}] \sim R_{yt}, \quad [R_{xt}, R_{yt}] \sim R_{xy} \tag{3.2}$$

and so on.

The enhanced algebra is called the *Lorentz algebra*, $SO(3, 1)$. In practical terms, there is a very big difference between $SO(3)$ and $SO(3, 1)$. Special relativity is equivalent to saying that the Lorentz algebra is a symmetry of nature. The classical phenomena associated to it are well-known: time dilation and length contraction, to name a few.

Here we will instead highlight physical effects in *quantum* theory. In this context Lorentz invariance is extremely restrictive compared to the rotation algebra, and leads to a number of important physical phenomena.

(i) *The number of polarisations of the photon*

The photon, like (almost) all particles in nature, has an intrinsic angular momentum or *spin*. In non-relativistic quantum mechanics we learn that an object of spin J (in units of Planck's constant $\hbar$) has $2J + 1$ independent polarisations. But the photon is fundamentally relativistic, after all it travels at the speed of light! And it does *not* satisfy the above formula. Rather, it has $J = 1$ but, being massless, only 2 polarisations:

$$\text{spin up} \uparrow \qquad \text{spin down} \downarrow$$

If there were "three polarisations of light", very basic phenomena involving transmission, reflection, diffraction and interference of light would be very different indeed!

(ii) *The existence of anti-particles*

The electron is known to have spin $J = \frac{1}{2}$ (again in units of $\hbar$). Dirac's famous wave equation for the electron was initially designed to be Lorentz invariant and describe electrons. But Lorentz invariance requires that every charged particle has an oppositely charged counterpart called its *anti-particle*.

There was absolutely no evidence for anti-particles in 1928 when Dirac proposed his equation. However, in 1931 he wrote another paper in which, based on his equation, he explicitly predicted the existence of an "anti-electron" with definite properties. Clearly he felt that if such a particle was required for consistency of his equation then it must exist in nature. And indeed, a year later a particle (the "positron") with precisely the predicted properties was discovered.

---

[2]due to Curt Rostenbach.

Figure 3. Paul Dirac.

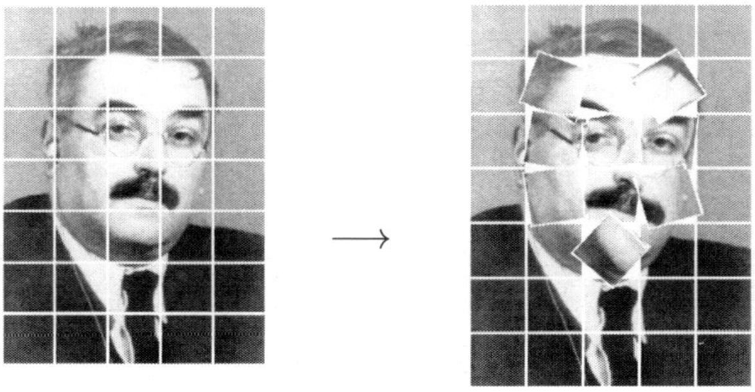

Figure 4. Rotating pieces of Dr. Fock independently!.

3.2 *Local version*

A remarkable generalisation of Lorentz symmetry is called "local Lorentz symmetry". This is a transformation like a rotation or boost, but the amount of rotation or boost is different at different points of spacetime.

We can illustrate this by taking a picture and rotating each tiny part of it independently. The result looks very different indeed from the original. Again, we cannot reasonably expect a picture to be invariant under such a bizarre operation, but it turns out that underlying theories of nature do have this invariance, which plays a deep and fundamental role as we now discuss.

(iii) *The existence of gravity*

Remarkably, local Lorentz invariance takes us from special relativity to general relativity, the theory of gravity. The very existence of a fundamental force – gravity – is predicted by this symmetry! This happens as follows. When we try to write down a mathematical theory that describes particles interacting via forces (such as a theory of electrons and photons), and then try to impose local Lorentz invariance on it, we find that this symmetry fails to hold because of some "extra terms" in the equations. To cancel out these extra terms and thereby restore the symmetry, a new dynamical variable has to be introduced. This new variable turns out to precisely describe the force of gravity in Einstein's general theory of relativity[3].

[3]The defining symmetry of general relativity is usually considered to be "general coordinate invariance", which is invariance under arbitrary (smooth) transformations of the space-time coordinates, but in more general contexts it is better to reformulate the theory as a local Lorentz invariant one.

As mathematicians know very well, "general coordinate invariance" is not really a symmetry at all. It is merely a redundancy in the description! The underlying geometry simply does not know about coordinates. We introduce coordinates as a crutch to help us work with the geometry, and then invoke general coordinate invariance to ensure that the answers do not depend on our choice of coordinates. This point is in fact very generally applicable. While "global symmetries" (where the parameter of the symmetry transformation is constant over space and time) are true symmetries, "local symmetries" (where the parameter varies arbitrarily over space and time) are redundancies of the description. Here we discuss both types of "symmetries" since they both have profound consequences for physics.

4. Internal symmetries

4.1 *Global version*

Let us now look at internal symmetries, which have no action on space or time. Instead, they rotate vectors in abstract "internal" spaces. An early example was isospin [4]. We think of

$$\begin{pmatrix} \text{proton} \\ \text{neutron} \end{pmatrix} \tag{4.1}$$

as a vector, and allow ourselves to perform rotations on it:

$$\begin{pmatrix} \text{proton} \\ \text{neutron} \end{pmatrix} \to \begin{pmatrix} U_{11} & U_{12} \\ U_{21} & U_{22} \end{pmatrix} \begin{pmatrix} \text{proton} \\ \text{neutron} \end{pmatrix} \tag{4.2}$$

where U_{ij} are the elements of a unitary matrix: $U^\dagger U = 14$.

This (approximate) symmetry predicts similarities in the behaviour of neutrons and protons, e.g. they have almost the same mass. This is nice, but not particularly surprising, as it's only what we expect a symmetry to do. Therefore one wouldn't say that isospin symmetry predicts any particular "phenomenon".

However, internal symmetries do give rise to physical phenomena under certain conditions. With standard assumptions, the most general internal symmetry is a combination of one or more compact Lie algebras. Let us denote the generators by T^a, where a runs over the dimension of the algebra. They obey the bilinear relation:

$$[T^a, T^b] = f_c^{ab}\, T^c \tag{4.3}$$

where f_c^{ab} are the structure constants of the algebra. The bracket above is an abstract (antisymmetric) pairing. For the Lie algebra to generate transformations of a physical system, the generators must act in definite representations. Now they become matrices, and the above bracket becomes a commutator. If at least one of the structure constants $f_c^{ab} \neq 0$, the algebra is non-Abelian.

We can now exhibit a phenomenon predicted by internal symmetries. We already saw in the above (also in Prof. Mukunda's article) that in a system with a symmetry, the states are typically not invariant. However the state of lowest energy (the vacuum) is special in that, depending on the theory, it may or may not be invariant under the symmetry. When the vacuum fails to be invariant under a continuous symmetry of the system, we have the phenomenon of "spontaneous symmetry breaking". An example from classical physics is shown in the figure.

(iv) *Goldstone bosons*

In quantum theory a spontaneously broken symmetry gives rise to the Goldstone phenomenon:

When a continuous symmetry is spontaneously broken, there must be a mode capable of rolling around in the ground state at no cost in energy.

Such a mode, in the context of quantum field theory, corresponds to the existence of a massless scalar (spinless) particle, the Goldstone boson. This severely constrains models of elementary particles, since no massless scalar particles are seen in nature.

4.2 *Local version*

Let us now consider the local version of internal symmetries. Here, one rotates the fields of elementary particles using matrices in a representation of some Lie algebra, but independently at each space-time point. Again this is should be thought of as a redundancy of the description rather than an invariance of a physical system. (It was Vladimir Fock who in 1926 first noticed that local gauge invariance was a property of Maxwell's equations.)

(v) *The existence of gauge particles*

As with the case of local Lorentz invariance, a new phenomenon emerges when we have local internal symmetries. Requiring the equations to be invariant under local transformations again gives rise to "extra terms". This time, to cancel them requires a spin-1 particle to exist for each generator of the Lie algebra. So just as local Lorentz invariance "predicts" gravity, local internal symmetries ("gauge symmetries") predict forces mediated by spin-1 particles.

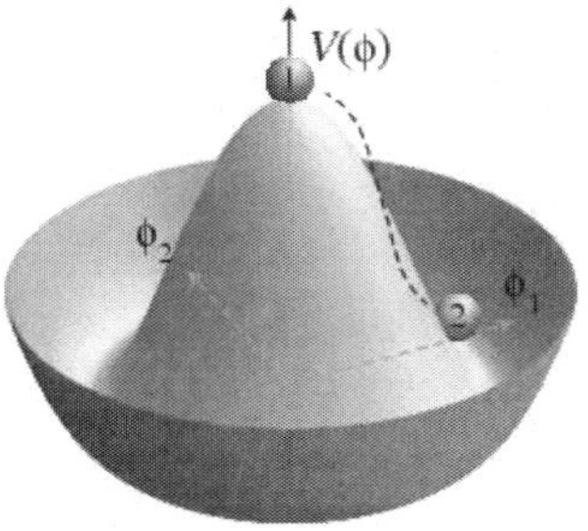

Figure 5. A particle at the top rolls down to its lowest energy state, but any point it rolls to is not invariant under rotations!.

Let us demonstrate in this case just how the force arises by cancelling "extra terms" associated to space-time dependent parameters. In this example we take the Abelian symmetry $U(1)$. This acts on the wave function of the electron by a phase rotation:

$$\delta\psi(\vec{x}, t) = i\alpha(\vec{x}, t)\, \psi(\vec{x}, t) \tag{4.4}$$

The energy associated to such a wave function typically involves derivatives, which change as:

$$\delta\frac{\partial\psi}{\partial x_\mu} = i\alpha\frac{\partial\psi}{\partial x_\mu} + i\frac{\partial\alpha}{\partial x_\mu}\psi \tag{4.5}$$

The energy function will not be invariant unless we cancel the second term, which involves derivatives of the symmetry transformation parameter α (and therefore would be absent if α were a constant as in global symmetry). This is done by introducing a spin-1 "gauge field" A_μ such that:

$$\delta A_\mu = \frac{\partial\alpha}{\partial x_\mu} \tag{4.6}$$

This is not just an example but a situation of physical relevance. The gauge field A_μ in this case is the photon. So with hindsight, we can say that the Maxwell equations (whose quantisation is the theory of Quantum Electrodynamics) could have been arrived at by postulating invariance under local Abelian phase transformations on the electron wave-function as a basic property of nature.

Like their global cousins, it seems that local internal symmetries can be spontaneously broken. However, we have argued that local symmetries are really redundancies. How can a redundancy possibly be broken? The equations resolve this paradox. While a broken global symmetry would have given rise to a Goldstone boson, with local symmetry the gauge field consumes it and changes its own properties, acquiring a *mass*. An immediate consequence is that forces that would have been long-range (massless particles can propagate arbitrarily far) become instead short-range, being attenuated by the effect of the mass.

Such massive spin-1 particles, associated to the "spontaneous breaking" of a local symmetry based on the Lie algebra $SU(2) \times U(1)$, have been discovered in 1983 at the Super Proton Synchrotron at CERN in Geneva, and are known as W^+, W^- and Z^0[4]. These three particles are the carriers of the weak nuclear interactions, which are indeed short-ranged [4].

(vi) *The Higgs effect*

The phenomenon by which a local symmetry that is "spontaneously broken" gives rise to massive gauge particles and thereby short-ranged forces, is called the Higgs effect. Because the spontaneous breaking needs to be seeded by some set of scalar fields that generate a potential (similar to the one in Figure 5), it often happens that one or more (massive but spinless) particles are left over at the end. These are called Higgs bosons.

[4] $SU(2) \times U(1)$ has four generators: there is therefore a fourth particle, which remains massless and is identified with the photon.

Figure 6. Large Hadron Collider.

At least one Higgs boson is required in the currently most successful theory of elementary particles. The search for these particles is a prime motivation for the Large Hadron Collider in Geneva.

We see that the cost of local internal symmetry can be as high as 9 billion dollars!

5. Supersymmetry

5.1 *Global version*

The idea that all internal symmetries are related to Lie algebras was generalised, in the 1970's, to include transformations involving the so-called *super-algebras*. These are known as *supersymmetry transformations*. Instead of transforming wave functions of spin-$\frac{1}{2}$ electrons into themselves, this symmetry transforms them into wave functions of scalar (spinless) particles:

$$\delta\psi \sim \phi, \quad \delta\phi \sim \psi, \tag{5.1}$$

Alternatively it can transform spin-1 gauge fields into neutral (uncharged) spin-$\frac{1}{2}$ fermions:

$$\delta\psi \sim A_\mu \quad \delta A_\mu \sim \psi, \tag{5.2}$$

In each case, supersymmetry changes the spin by $\frac{1}{2}$ in the standard units.

The reasons for invoking supersymmetry are related to the dynamics of unification of forces [4], which we will not discuss here. However it is quite evident that the existing particles do not fall (even approximately) into pairs under supersymmetry. For example there is no known massless fermion that could be the partner of the "photon", and it certainly would have been detected if it had the same type of interactions as the photon which is what supersymmetry predicts.

Therefore supersymmetry, if true, requires a whole new set of particles: one new particle as the super-partner of every known particle, and moreover each new particle must be "split" in mass from the original one in order to have escaped detection. This can happen if supersymmetry is an approximate or "broken" symmetry, which is the current belief held by a number of physicists[5].

[5]Of course the alternative is that, despite pressing motivations related to unification, supersymmetry is actually not realised in nature.

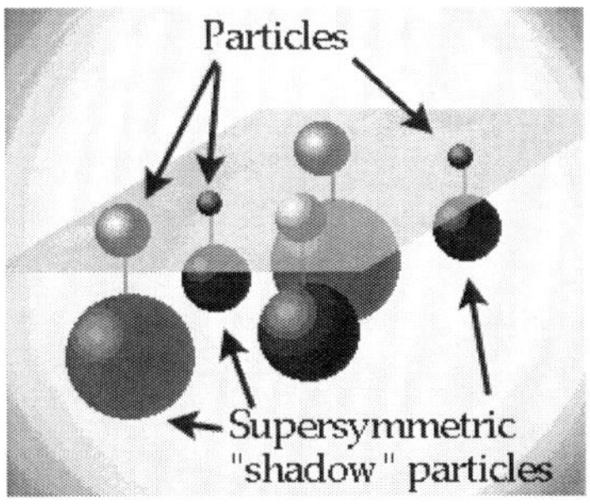

Though supersymmetry is an internal symmetry, it is linked to space-time. Indeed it acts as a transformation in a generalised space-time called "superspace" with commuting and anti-commuting directions[6]. Moreover, the anti-commutator of two supersymmetry transformations is an ordinary translation in space-time.

$$\{S, S^{\dagger}\} \sim T \tag{5.3}$$

This is the first time in physics that a basic symmetry like translation has been written as a composite of another symmetry!

Like other Lie-algebra-based symmetries, supersymmetry can be spontaneously broken. In this case, an analogue of Goldstone's theorem predicts a massless fermionic particle called a Goldstino.

5.2 *Local version*

(vii) *The existence of supergravity*

Since supersymmetry is not yet established, the previous discussion as well as the one to follow should be understood as being hypotheses about nature. Of course on the mathematical side supersymmetric theories are as well-defined as any others (some would even say "better defined"!).

As with Lie algebra symmetries, there is a local version of supersymmetry. Here a boson transforms into a fermion by an amount that is different at each point of spacetime. Now a beautiful thing happens. Since two supersymmetries combine into a translation, local supersymmetry implies *local translation symmetry*. This is the same as symmetry under arbitrary changes of coordinates at each point, namely general coordinate invariance.

This, as we have seen, implies gravity – or rather, a supersymmetric extension of it: called *supergravity*. The particle that mediates gravity is the spin-2 graviton, denoted $g_{\mu\nu}$. Local supersymmetry pairs this with a new particle, the spin-$\frac{3}{2}$ gravitino denoted χ_μ:

$$\delta\chi_\mu \sim g_{\mu\nu}, \quad \delta g_{\mu\nu} \sim \chi_\mu \tag{5.4}$$

The gravitino – yet another particle predicted by symmetry – is being sought (along with other "superpartners") at the LHC.

---

[6]Anti-commuting directions arise because particles of half-integer spin are fermions, described by anti-commuting wave-functions, and supersymmetry has to be fermionic in order to change the spin of a particle by $\frac{1}{2}$.

6. 3-algebras

In recent years a new type of symmetry has attracted interest. It originated, like many other good things, with Prof. Yoichiro Nambu. His idea was to generalise the 2d phase space, which comes with variables (p, q), to a 3d space with variables, say, (p, q, r). Then the Poisson bracket can be naturally generalised:

$$[G, H] \to [F, G, H] = \frac{\partial F}{\partial p}\frac{\partial G}{\partial q}\frac{\partial H}{\partial r} \pm \cdots \tag{6.1}$$

The idea was revived recently by Basu and Harvey [5] and by Bagger and Lambert [6] in connection with an open problem: how to find a maximally supersymmetric, scale-invariant field theory in $(2 + 1)$ dimensions. A rather special feature of this theory is that it is uniquely determined by supersymmetry. This field theory plays an important role in describing membrane excitations in a strongly-coupled limit of string theory called M-theory, which we will not go into here. But its existence was conjectured over a decade ago. In $(2+1)$ dimensions it is known that scalar fields have a scale dimension $\frac{1}{2}$ and therefore the interaction ϕ^6 is dimensionless – a necessary (though not sufficient) requirement for scale invariance. This fact will play a role in formulating the desired field theory.

The novel mathematical structure employed in these works is the concept of a 3-algebra:

$$[T^a, T^b, T^c] = f_d^{abc}\, T^d \tag{6.2}$$

generalising the notion of Lie algebra:

$$[T^a, T^b] = f_c^{ab}\, T^c \tag{6.3}$$

If we have a number of scalar fields ϕ^{Ia} and we write:

$$\boldsymbol{\phi}^I = \phi^{Ia} T^a \tag{6.4}$$

then it is natural to postulate a 6-th order coupling:

$$\mathrm{tr}[\boldsymbol{\phi}^I, \boldsymbol{\phi}^J, \boldsymbol{\phi}^K]^2 \sim f_g^{abc} f^{defg} \phi_a^I \phi_b^J \phi_c^K \phi_d^I \phi_e^J \phi_f^K \tag{6.5}$$

The 3-algebra symmetry is similar to a local gauge symmetry in that it requires a gauge-field, which however is non-dynamical and does not correspond to a new force.

In this theory it has been shown [7] that the Higgs mechanism acts in a novel way. It renders the gauge field dynamical and reduces the theory to a conventional non-Abelian gauge theory – an important consistency condition arising in string theory.

The role of 3-algebras in physics is not yet definitively established. It will be interesting to see how this develops.

7. Conclusions

We have seen that Lie algebras and super-algebras have a predictive power: The presence of symmetry implies a (frequently unexpected) physical phenomenon. It is deeply satisfying that nature uses symmetry principles to determine its dynamics! New symmetries may yet be found and solve our remaining puzzles. The ongoing story of 3-algebras is a possible example.

References

[1] For some details and references, see: http://en.wikipedia.org/wiki/Group_(mathematics).
[2] For an introductory review of string theory, see: http://theory.tifr.res.in/~mukhi/Physics/string2.html
[3] N. Mukunda, The Three Roles of Symmetry in Fundamental Physics, this volume.
[4] A review of contemporary particle physics, which explains many of the concepts discussed in the present article, is: S. Mukhi and P. Roy, Developments in high energy theory, *Pramana*, **73** (2009) 3 [arXiv:0905.1793 [physics.pop-ph]].
[5] A. Basu and J. A. Harvey, The M2-M5 brane system and a generalized Nahm's equation, *Nucl. Phys. B*, **713** (2005) 136 [arXiv:hep-th/0412310].
[6] J. Bagger and N. Lambert, Gauge Symmetry and Supersymmetry of Multiple M2-Branes, *Phys. Rev. D*, **77** (2008) 065008 [arXiv:0711.0955 [hep-th]].
[7] S. Mukhi and C. Papageorgakis, "M2 to D2", *JHEP*, **0805** (2008) 085 [arXiv:0803.3218 [hep-th]].

Symmetry: A Multi-Disciplinary Perspective No. 16, 2011, pp. 25–34.

The Three Roles of Symmetry in Fundamental Physics

N. Mukunda

103, 6th Main Road Malleswaram, Bangalore 560 003, India
e-mail: nmukunda@gmail.com

Abstract. We review the three major roles of symmetry in the development of physical theories, namely the descriptive, restrictive and creative roles.The concepts of events, laws of Nature, and their symmetries in analysing experience are emphasised. Examples from Galilean relativistic physics, from special relativity, from classical and quantum domains are presented.

Keywords. Symmetry, physical laws, relativity, classical mechanics, quantum mechanics.

1. Introduction

It is quite well known and widely appreciated that the concept of symmetry is of fundamental importance in physics. I will describe aspects of this connection in a historical perspective. The perfect conceptual framework for capturing and expressing the idea of symmetry in relation to natural phenomena is that of a group. Indeed it has been said that this is the greatest gift of 19th century mathematics to physics.

In 1951, on the eve of his retirement from the Institute for Advanced Study in Princeton, Hermann Weyl gave a set of four lectures titled "Symmetry" which were published in 1952 and have since become justly famous. Here is a beautiful sentence from the early part of this book:

"Symmetry, as wide or as narrow as you may define its meaning, is one idea by which man through the ages has tried to comprehend and create order, beauty and perfection."

In his lectures, Weyl mainly considers what can be called the static case, the geometric symmetries of objects and their arrangements in three dimensional space. His examples are drawn from living organisms, architecture, sculpture and painting. In all these it is finite and discrete groups which play an important role. We will go well beyond that treatment in two respects – on the one hand we consider continuous or Lie groups as symmetry groups; on the other hand we look at their actions on physical systems involving time in an essential way. This means that we are interested in dynamical symmetry rather than three-dimensional spatial or geometric symmetry. This is a level of symmetry which "cannot be seen by the eye but only by the mind."

Invited inaugural talk at the 'Symposium on Symmetry', Indian Institute of Science Education and Research, Mohali, 19–20 February, 2010.

Weyl does mention this but only very briefly at the end of his discussion. Naturally a fair degree of abstraction and sophistication is needed here, which I will try to explain as we go along.

2. Events – Laws of Nature – Symmetries

At several places in his general writings Eugene Wigner has emphasized that for the description and analysis of physical phenomena we need three key concepts arranged in a natural hierarchy. The first and most primitive or elementary one is that of an "event". This is usually pictured as something that "happens" at a certain place at a definite time. It could be the passage of a point particle through a spatial location at some time; a collision between two particles; the value of a field component at a space time point; the emission of a flash of light, etc. We can generalise this notion and think of an "event" as the state of a physical system at a certain time. How this is to be described depends upon the system considered and also whether we are working in the classical or quantum framework.

The next concept in the hierarchy is that of a history, a sequence of events in time, which is physically permitted or can be realised. Such histories are selected out of all imaginable histories by definite equations of motion which can be expressed via an Action Principle. Typically they are systems of ordinary differential equations or of partial differential equations, examples being the Newtonian equations of mechanics, Maxwell's equations for electromagnetism, the Schrödinger wave equation in quantum mechanics, etc. Stated briefly, allowed or possible histories are solutions of the equations of motion, which in turn are the Laws of Nature. As Wigner says:

"The purpose of all equations of physics is to calculate, from the knowledge of the present, the state of affairs that will prevail in the future."

So in passing from individual events to individual strings of events or histories and then on to all possible histories we have a great contraction or compression of the contents of immediate experience. A definite physical system being given, its infinitely many realisable histories are all seen as being solutions to a common system of equations of differential type, a common dynamical law. And one then has to develop the ability to deal with the set or space of all solutions of a given system of equations of motion efficiently.

The second and last step in this context takes us from equations of motion, or laws of nature, to their symmetries. As we shall see, this often leads to a further contraction or unification of elements of experience – many laws of nature may be governed by the same group of symmetry transformations. It is important to realise that the symmetry transformations act on the basic variables of the system and possibly on time too in such a way that the equations of motion, or dynamical laws, are preserved. They are not symmetries of a particular state or of the condition of the system at a specific time. This is what makes such symmetries dynamic rather than static. Briefly and for emphasis – it is the **equations** determining development in time that are preserved, not an **individual state** at a particular time nor even a **particular history**. Thus each symmetry operation maps each solution of the equations of motion generally onto some other solution. It acts as a map on the space of all possible states of motion in an invertible and consistent manner.

This role of symmetry – what it is and what it does – is clearly quite refined and abstract. It is what makes symmetry "visible" only to the "mind's eye," as I said earlier. Its importance is well conveyed by Valentine Bargmann in these words:

"... those laws of physics which express a basic 'invariance' or 'symmetry' of physical phenomena seem to be our most fundamental ones."

So we have three levels of successively increasing depth and abstraction – events, laws of nature determining possible strings of events or histories, symmetries of the laws often common to many. In some of his writings Wigner uses the words "initial conditions" rather than "events" in this three-fold pattern, the two are essentially the same though "event" seems preferable. One point he makes in this connection is quite important – he says it was Newton's genius that made him carefully distinguish between initial conditions and the laws of motion, which is where regularities, symmetry and universality are all seen. Quoting from him:

"The sharp distinction between Initial Conditions and Laws of Nature was initiated by Isaac Newton and I consider this to be one of his most important, if not **the** *most important accomplishment. Before Newton there was no sharp separation between the two concepts. After Newton's time the sharp separation of initial conditions and laws of nature was taken for granted and rarely even mentioned."*

3. The three roles of symmetry

Echoing this three-level structure we will see next that symmetry too can operate – or has historically operated – in three ways: descriptively, restrictively and creatively. The meanings of these terms, these roles of symmetry, are best brought out by recalling examples of each.

3.1 *Symmetry as descriptive*

It is well known that the equations of mechanics in the Galilean-Newtonian framework are invariant under a particular space time transformation group, namely the Galilei group. The special Galilean or velocity transformations have the familiar rather simple form

$$\underline{\mathbf{x}}' = \underline{\mathbf{x}} - \underline{\mathbf{v}}t, \quad t' = t \tag{1}$$

in going from one inertial frame to another with a relative velocity $\underline{\mathbf{v}}$. Apart from these, the group contains three-dimensional rotations, and translations in space and time. However, an understanding of these invariances of the equations of non-relativistic mechanics came much later than the discovery of the equations themselves, so they became a description of properties of already known physical laws. In Newton's time and for long thereafter, the concept of a group did not exist. Indeed a full appreciation of the role of Galilean invariance came by a reexamination of classical mechanics after the advent of quantum mechanics.

In the case of Maxwell's equations for classical electromagnetism the situation is in some ways similar but with dramatic differences as well. Here again the (field) equations were found first, in 1865, while their symmetry group was finally understood about four decades later with the Lorentz transformations of special relativity. In this Lorentz and Poincaré played very important roles. Here the relation between two inertial frames with relative velocity $\underline{\mathbf{v}}$ is given by the more intricate pure Lorentz transformations which replace (1) above:

$$\underline{\mathbf{x}}' = \frac{\underline{\mathbf{x}} - \underline{\mathbf{v}}t}{\sqrt{1 - \underline{\mathbf{v}}^2/c^2}}, \quad t' = \frac{t - \underline{\mathbf{v}} \cdot \underline{\mathbf{x}}/c^2}{\sqrt{1 - \underline{\mathbf{v}}^2/c^2}} \tag{2}$$

The replacement for the Galilei group here is now the Poincaré group, which contains spatial rotations and space time translations in addition to the pure Lorentz transformations. But it turned out that these Lorentz transformations express intrinsic properties of space and time, though they were first seen through Maxwell's equations. This is the point at which the transition to another role of symmetry took place.

3.2 *Symmetry as restrictive*

It was in the opening years of the last century, the early 1900s, that a new point of view emerged regarding the role of symmetry in relation to physical law. This was due principally to Poincaré and Einstein, the idea being not to regard symmetry as an afterthought but as a starting requirement. The motivation was the realisation that the Lorentz transformations disclosed by classical electromagnetism have an applicability beyond that domain, as they embody the very structure of space-time which is the stage for all physical phenomena. As we know, Einstein modified Galilean invariant Newtonian particle mechanics so as to be Lorentz invariant and thus fall in line with electromagnetism.

Two statements, from Einstein and from his one-time assistant Valentine Bargmann whom I quoted earlier, convey this restrictive role of the symmetry of special relativity very well:

"The Principle of Relativity is a principle that narrows the possibilities; it is not a model, just as the Second Law of Thermodynamics is not a model."

"... every physical theory is supposed to conform to the basic relativistic principles and any concrete problem involves a synthesis of relativity and some specific physical theory."

Let me now present several examples of the results of imposing Lorentz symmetry on physical theories in various domains. You will see this role of symmetry in splendid action. Remaining faithful to the principle of three, I shall give you three examples each from classical physics, the Old Quantum Theory, quantum mechanics and quantum field theory.

Restrictive role of symmetry

- Mass Energy Equivalence $E = mc^2$
- Ten Conservation Laws
- Dirac-Lorentz Equation

- Sommerfeld Fine Structure Formula
- Photon Momentum $p = E/c$
- Planck's $E = h\nu$ to de Broglie's $p = \hbar k$

- Dirac Electron Equation
- Weyl Neutrino Equation
- Wigner Analysis of Elementary Systems

- Fermi Weak Interaction Theory
- Pauli Spin Statistics Theorem
- Tomonaga Feynman Schwinger Renormalization Theory

Perhaps the best known classical result is the mass energy equivalence relation $E = mc^2$. This was found by Einstein by amending Galilean Newtonian particle mechanics, as mentioned earlier, to share the Lorentz invariance of electromagnetism. In turn this relation united the earlier separate conservation laws of energy and of mass into one law. Recall that Galilean symmetry leads to both these laws, in an extremely subtle manner.

Beyond this, Lorentz invariance of any classical theory based on an Action Principle automatically leads to the ten mechanical conservation laws of energy, momentum, angular momentum, and moment of energy. The first statements of this important connection are by G. Hamel in 1904, and E. Noether in 1918. The most spectacular use of this is in Dirac's 1938 treatment of the classical relativistic point electron – using just the energy momentum conservation laws and manifest Lorentz covariance he was led to the famous Dirac-Lorentz equation which included the effects of radiation reaction. In a sense this is the ultimate classical theory of charged relativistic point particles. It is interesting to remember that this work of Dirac was followed up by J.A. Wheeler and R.P. Feynman at Princeton University in the USA, and by H.J. Bhabha and Harish Chandra at the Indian Institute of Science at Bangalore in the early 1940's.

Turning to the period of the Old Quantum Theory from 1913 to about 1924, we have first Sommerfeld's derivation of the fine structure of the hydrogen spectrum as a relativistic correction to Bohr's theory, but later developments showed that this was fortuitous. Then came Einstein's association in 1916 of a momentum $p = \hbar k$ to a light quantum with energy $E = h\nu$, the two being connected by $E = pc$. This depends upon and can only be understood on the basis of special relativity. So these two Einstein relations $E = h\nu$ and $p = E/c$ for photons were separated in time by eleven years! A few years later in 1923 de Broglie was guided by special relativity in extending the Planck relation $E = h\nu$ to the momentum wave number relation $p = \hbar k$

for material relativistic particles. This led to his associating a relativistic wave with a moving particle, the starting point of wave mechanics.

Now we come to quantum mechanics proper, set up in 1925–26 by Heisenberg, Dirac and Schrödinger, more particularly to its relativistic extension in 1928 through Dirac's four component wave equation. The ingredients that led to this truly stunning development were the general structure of quantum mechanics, the requirement of symmetry with respect to the Lorentz group, and of course Dirac's unique genius. Many successes came out of this equation – understanding of electron spin, its anomalous magnetic moment, the fine structure of the hydrogen spectrum, and finally the positron and the general idea of antimatter. Actually the prediction of the positron was contained in his 1931 paper on the magnetic monopole in quantum mechanics. Soon after Dirac's initial successes, even before Pauli had proposed the existence of the neutrino in 1930, Weyl in 1929 had proposed a two component relativistic wave equation for particles with spin one half and zero mass. Today we know there are three neutrinos and at least two of them have nonzero mass, so after so many decades the Weyl equation has diminished usefulness.

The third item in this list of quantum mechanical consequences of relativistic symmetry is Wigner's construction of unitary irreducible representations of the Poincaré group. Started around 1930 upon Dirac's suggestion and completed in 1937, this was a major achievement both for physics and for mathematics, and it led to a classification of all possible elementary particles – or elementary systems as Wigner often called them – based on the two characteristics of mass and spin, the two invariants of the Poincaré group.

Last in this account of symmetry as restrictive are three instances from quantum field theory. In 1934 Fermi created a particular theory of the weak interactions responsible for beta decay. It soon became clear that on the basis of Lorentz invariance and parity conservation there were five independent forms for this interaction. Then in 1956 Lee and Yang proposed that parity was not conserved in weak processes; and then the number of possible Lorentz invariant interactions increased to ten. But it was quickly reduced to just one in 1957 by Sudarshan and Marshak with their universal V-A interaction incorporating chirality invariance.

Within quantum field theory itself the connection between spin and statistics was shown by Pauli in 1940 to be a consequence of relativity. In fact he ends his paper with the sentence:

"*. . . we wish to state, that according to our opinion the connection between spin and statistics is one of the most important applications of the special relativity theory.*"

However E.C.G. Sudarshan has argued that this connection depends only on three-dimensional rotational invariance and the Action Principle.

Later in the 1940s manifest Lorentz invariance was a crucial ingredient for Tomonaga, Feynman and Schwinger to develop renormalization theory to handle divergences in quantum electrodynamics and get meaningful predictions for comparison with experiment.

3.3 *Symmetry as creative*

It is by taking the restrictive role of symmetry to its logical conclusion in various contexts that we arrive at the final creative role. In fact one can say this is the deepest and the most sophisticated functioning of symmetry. The idea is that given the physical domain one is interested in and a definite set of transformations as symmetry group, the restrictive power is so great that eventually only one theory survives. There are some technical aspects involved here, but I will try to convey the essential ideas. It should be no surprise that we have three examples to consider.

Creative role of symmetry

- General coordinate transformation invariance $\longrightarrow$ General Relativity
- Abelian gauge invariance $\longrightarrow$ Electrodynamics
- Non-abelian gauge invariance $\longrightarrow$ Yang Mills Theory

The original way in which the equations of general relativity were found depended very heavily on the invariance requirements placed upon them. These turned out to be strong enough to determine the equations, of course aided by deep physical insight. One considers events taking place in space and time and labels them using space time coordinates. Einstein assumed that space time is a pseudo Riemannian manifold, with complete freedom in the assignment of coordinates to events:

$$\text{Events in space time} \longrightarrow \text{coordinates } x^{\mu}, \qquad \mu = 0, 1, 2, 3;$$

$$\left.\begin{array}{l}\text{arbitrary (smooth) general}\\ \text{coordinate transformations}\end{array}\right\} : x^{\mu} \longrightarrow x'^{\mu} = \text{functions of } x \tag{3}$$

Here one is not limited to transformations like those of Galilean relativity or special relativity, as the concept of preferred global inertial frames is given up. Thus one has the freedom to carry out arbitrary general coordinate transformations, and the equations one is searching for must preserve their forms under these transformations. The basic physical insights were the expression of the Principle of Equivalence and the identification of gravitation with space time curvature. With the (pseudo)Euclidean geometry of special relativistic space time no longer available, the metric tensor of space time became the field representing gravitation. When one has the freedom to carry out essentially arbitrary general co-ordinate transformations (3), both general tensorial field quantities and covariant differentiation processes have to be set up; these are provided by the methods of the absolute differential calculus developed after Riemann by a galaxy of geometers. It is with these ingredients that one has a systematic procedure to construct differential equations of motion that preserve their forms under the transformations (3). These processes have to be applied to the metric tensor field to generate second order differential field equations, consistent with local energy momentum conservation and reducing to Newtonian Theory in the appropriate limit. Einstein was able to solve this problem, and general relativity was formulated in 1915.

Very soon after, in 1918, Weyl proposed a truly beautiful extension of Einstein's theory, which seemed capable of unifying gravitation and electrodynamics, the two fundamental interactions known at that time. This involved a generalisation of Riemannian geometry to what is called Weyl geometry. To the freedom of performing general coordinate transformations, Weyl added a new group of symmetry transformations – the freedom to change the scale of length (and so of time as well) independently at each space time point in a continuous manner. He called these "gauge transformations" and showed that to accomodate them a "vector potential" was needed, leading in turn to Maxwell's equations. However, Einstein pointed out that this proposal was untenable from the physics point of view, and this attempt of Weyl had to be given up.

The idea itself was however so beautiful that it would not go away. About a decade later, in 1929 after the arrival of quantum mechanics, Weyl, Fock and London independently realised that the idea of gauge transformations can be revived in a new context – instead of real local scale changes, they correspond to complex local phase changes of Schrödinger wave functions of electrically charged particles. This switch from the real to the complex domain occurred because in Schrödinger wave mechanics the classical real momentum variable is represented by i times a differential operator acting on wave functions. Recall also that electromagnetic interactions of charged particles are described by the minimal coupling rule $p \to p - (e/c)A$ in the classical Hamiltonian, and this is carried over to quantum mechanics. So real local gauge transformations were altered to local phase transformations, and the requirement of symmetry then led to electrodynamics in an acceptable fashion. The phrase "gauge invariance" was however retained.

In this case, the important non-space-time group involved is the group of phase factors which is the abelian unitary group $U(1)$ in one dimension. In 1954, Yang and Mills found an extension to general nonabelian compact Lie groups and set up what is called nonabelian gauge theory. This is again a structure of great beauty, standing in complexity in between general relativity and electrodynamics. The arbitrary space-time dependent transformations act in a multidimensional internal space describing properties which are a generalization of electric charge. Once again there is a vector potential which transforms in a more intricate way than in the case of electromagnetism, and the increased demands of symmetry practically determine the basic equations. Actually, already in 1939 Oskar Klein had come very close to this theory, but in the work of Yang and Mills it achieved a definitive form.

The occurrence of transformations with arbitrary space-time dependences, characteristic of all three creative uses of symmetry, has some extremely subtle technical consequences. The connection between continuous symmetries and conserved quantities, based on an action principle, always holds. However in these cases the conserved quantity is limited to the value zero, it becomes a constraint on dynamical variables. At the same time, such symmetry transformations are better understood as relating different descriptions of the same physical situation, not as leading from one physical state to another distinct one. These special features of such symmetry transformations must be kept in mind and understood with care.

For many years, the theory of Yang and Mills could not be used successfully in physics, because it required the presence of many massless particles analogous to the photon. This difficulty was finally overcome during the 1960s with the idea of spontaneous symmetry breaking, whose most gifted exponent is Yoichiro Nambu. Today nonabelian gauge theory with spontaneous symmetry breaking, in a quantum mechanical framework, is the basis for the unification of the weak and electromagnetic interactions; the current theory of the strong interactions is an independent pure Yang Mills theory.

So these are the three examples of symmetry in its finest creative role.

4. Concluding remarks

I have tried to give a pedagogical account of the three roles of symmetry in physics in a historical perspective, with selected examples. No doubt others might have been chosen. It is time now to conclude with a few remarks.

The importance of symmetry is greatly enhanced in the framework of quantum mechanics as compared with classical physics, because of the superposition principle and the Hilbert space setting. This makes the entire mathematical apparatus of unitary representations of Lie (and other) groups, more precisely their ray representations, directly relevant for physical problems in a very satisfying manner.

There have been some extensions of the Lie group concept suggested by developments in physics which should be mentioned. One is the idea of supergroups leading to super symmetry, involving transformations which connect bosons and fermions with one another and using Grassmann variables as group parameters. Another is the idea of quantum groups suggested by more specialised problems of physical interest. These are developments worth serious study.

I began by quoting from the opening pages of Weyl's book 'Symmetry'. It is fitting now to go to his closing pages:

'Symmetry is a vast subject, significant in art and nature. Mathematics lies at its root, and it would be hard to find a better one on which to demonstrate the working of the mathematical intellect.'

To this I would add that it is thanks to the physicist's intuition that such deep connections between the mathematics and the physics have surfaced. And I would point again to the important ways in which this account goes beyond the scope of Weyl's treatment.

Some references and suggestions for further reading

[1] H. Weyl, Symmetry, Princeton University Press, Princeton, New Jersey (1952).

[2] E. P. Wigner, Symmetries and Reflections, Indiana University Press, Bloomington, Indiana (1967); Ox Bow Press, Woodbridge, Connecticut (1979); Philosophical Reflections and Syntheses, Springer (1995).

[3] V. Bargmann, Relativity, *Reviews of Modern Physics*, **29** (1957) p. 161.

[4] The Collected Papers of Albert Einstein, Volume 3: The Swiss Years: Writings, 1909–11, Princeton University Press (1993); p. 357, Comments after a lecture on The Theory of Relativity in Zurich, 16 January (1911).
[5] G. Hamel, *Z. Math. Phys.*, **50** (1904) 1.
[6] E. Noether, Invariante Variationsprobleme, *Nachr. Kgl. Ges. Wiss. Gottingen*, **235** (1918) p. 37.
[7] P. A. M. Dirac, Classical Theory of Radiating Electrons, *Proc. Roy. Soc. London*, **A167** (1938) 148.
[8] P. A. M. Dirac, The Quantum Theory of the Electron I, II, *Proc. Roy. Soc. London*, **A117**, 610; **A118**, (1928) 351.
[9] E. P. Wigner, On unitary representations of the inhomogeneous Lorentz group, *Ann. Math.*, **40** (1939) 149.
[10] W. Pauli, The Connection Between Spin and Statistics, *Phys. Rev.*, **58** (1940) 716.
[11] W. Pauli, Theory of Relativity, Pergamon Press (1958).
[12] C. N. Yang and R. L. Mills, *Phys. Rev.*, **95** 631; **96** (1954) 191.
[13] O. Klein, On the Field Theory of Charged Particles, New Theories in Physics, International Institute of Intellectual Cooperation, Paris (1939); reprinted in The Oskar Klein Memorial Lectures, Vol. 1, Ed. Gosta Ekspong, World Scientific (1991).

Symmetry: A Multi-Disciplinary Perspective No. 16, 2011, pp. 35–48.

Symmetry in the Living World: How and Why

Vidyanand Nanjundiah

Indian Institute of Science and Jawaharlal Nehru Centre for Advanced Scientific Research, Bangalore 560 012, India
e-mail: vidya@ces.iisc.ernet.in

Based on a talk given at the Indian Institute of Science Education and Research, Mohali, on February 20, 2010.

Abstract. Symmetry is a pervasive feature of the living world. Leaves, butterflies, fish, and human bodies display approximate left-right symmetry; frog eggs show cylindrical symmetry; and adult sea urchins and starfish possess radial symmetry. Besides those, many embryos exhibit 'scale-invariance' to some degree: over a range of sizes, the relative proportions of the body parts (of the eventual adult) are more or less the same. This essay is meant to serve as an introduction to the theme of symmetry in biology and is written in an informal style, so that it can be of use to readers who do not have a background in biology. It focuses on two aspects of symmetry: how does it originate during individual development, and what might be the evolutionary reasons behind its existence. A brief mention is made of brain asymmetry. The mathematical basis of 'Turing' patterns is sketched in a self-contained Appendix at the end. The essay should be read while glancing at the illustrations that were screened during the talk; this can be done by going to the Web sites cited.

Introduction

Symmetry is an attribute that can be used to refer to abstract entities such as mathematical equations and physical laws as well as to tangible objects. As used by biologists, it may be defined as follows: a structure is symmetric if it looks approximately the same after it has been subjected to a transformation in space or time. The transformations that are relevant are the usual ones of displacement, rotation and reflection. If we exclude science fiction, only the first of them applies in the case of temporal symmetry. Symmetry (and its counterpart, asymmetry or, more familiarly, 'handedness') is ubiquitous in the living world and, as with much else, ought to be accounted for as an outcome of evolution. A type of symmetry may be found in a particular lineage because of specific evolutionary pressures in the history of that lineage.

The concept of symmetry does not appear to possess the same deep significance in biology that it does in physics. For example, one can hardly conceive of its being implicitly linked to a conservation law. It has been speculated that the handedness or chirality of the amino acids found in proteins may be linked to a breakdown of left-right symmetry in the weak force. All but one of the 20 common amino acids that make up proteins can exist – in principle – in one of two optically active isomeric forms, *L* or 'left-rotatory' and *D* or 'right-rotatory'. In practice, however, they were believed to be always found in the *L* form – until recently. It turns out that *D*-amino acids are found in some opioid peptides – molecules with analgesic properties – and snake venoms; also, they accumulate in the body with age (http://faculty.bennington.edu/~jbullock/Chem1/d%20amino%20acids.pdf). This essay is restricted to symmetry at the level of organisms; it does not deal with molecular asymmetry. A list of references is provided at the end for readers who may want to go deeper into the subject of biological symmetry.

The qualification 'approximate' made in the previous paragraph is necessary because symmetries in the living world almost never approach the degree of precision that is displayed by, for example, crystal structure. To put it loosely, biology is less precise than elementary physics. Typically, left-right asymmetry in a body part that is conventionally described as being symmetrical is of the order of 0.5–2.5% (see Møller and Thornhill, 1998); a figure of 5% would be quite high. In contrast, crystal lattice parameters (~ the spacing between successive planes) are said to be invariant within 1 part in 10^6 (Becker, http://iopscience.iop.org/0034-4885/64/12/206); the extent to which electromagnetic transitions in atoms do *not* conserve left-right reflection symmetry is quoted as 1 part in 10^{16} or less (Warrington et al., http://iopscience.iop.org/1402-4896/1995/T59/061). Exceptions arise in the case of microscopic and sub-microscopic life, for example diatoms and viruses. The reason is that the forms of very tiny creatures are, so to say, one-step and direct consequences of physical principles similar to those that underlie crystallisation. Molecular transport and dispersed chemical reactions become significant with an increase in size; the integration of the whole, moulded by natural selection, becomes of overwhelming importance at still larger scales. All these factors contribute to departures from ideal symmetry. To put it differently, if 10^4 components are assembled independently, and each of them is specified within a tolerance of 1 in 10^6, the overall tolerance can be specified to no better than 1 part in 100. A macroscopic physical structure consisting of very many microscopic sub-structures would be expected to show a comparable level of imprecision.

Biological symmetries

Prominent symmetries exhibited by living forms are mirror-image reflection (bilateral symmetry), rotation, and rotation combined with translation (helical or spiral symmetry). The second category includes cylindrical symmetry, that is, rotation about an axis through an arbitrary angle, and 3-fold, 4-fold, 5-fold, 6-fold, 8-fold or higher rotational symmetry, going all the way to near-spherical symmetry (http://en.wikipedia.org/wiki/File:Haeckel_Discoidea.jpg). The presence of 5-fold symmetry (forbidden by crystallography) shows that we are dealing with biology, not physics. A multi-protein structure, the proteasome, has been found to exhibit 7-fold symmetry.

There are unexpected features of symmetry in biology. First, the type of symmetry can change during the lifetime of an individual (e.g. a cylindrically symmetric sea urchin larva gives rise to a bilaterally symmetric adult; a bilaterally symmetric flatfish becomes asymmetric; a left-handed lobster can become right-handed following the loss of an appendage and regeneration; see Ophiopluteus larva and adult, http://life.bio.sunysb.edu/marinebio/larvae.html). Second, an individual can exhibit more than one form of symmetry, sometimes in different parts (e.g. a plant can have both rotationally symmetric and bilaterally symmetric flowers; a fly larva can display bilateral symmetry as well as translational symmetry along its length). Third, symmetry may exist only at the level of the population, with each individual being either left- or right-handed (as in some flatfish).

A sub-category of population-level bilateral symmetry is known as *fluctuating asymmetry*. This means that although individuals are on the whole bilaterally symmetric, each of them displays a slight bias towards right- or left-handedness. Often the bias cannot be ascribed to heredity (i.e., to genetic factors) but is an offshoot of the vagaries of development (the numbers of tail feathers on the left and right sides of the body in birds provides one instance).

Allowing for differences in sex, the size reached by an adult is among the defining features of living systems. It comes as a surprise, therefore, that many biological patterns can be indifferent to overall size (scale invariance). A familiar example is when a single fertilised egg splits into two and gives rise to identical twins (properly referred to as monozygotic twins), sometimes of opposite handedness (for a frog example see http://www.nature.com/nrm/journal/v7/n4/fig_tab/nrm1855_F1.html). More spectacular cases involve the production of identical quadruplets or octuplets, as for example in the case of the armadillo, where this happens routinely (see http://www.valdosta.edu/~jloughry/Reprints/AmSci.pdf). In contrast to scale-invariance, there are many examples of *allometric* development. If we denote the linear dimensions of two body parts by x and y, scale invariance implies that x is proportional to y; in allometry, x is proportional to a power of y other than 1 (see http://en.wikipedia.org/wiki/Allometry).

One can also detect temporal symmetries in living forms. The most obvious of them relates to the reiteration of early cell lineages during the development of a fertilised egg. In other words, not only do cell divisions occur at a regular rate; it appears that a 'daughter' cell can perform the same role as its mother. Also, oscillatory phenomena at the level of cells, tissues or whole organisms – are common in biology.

Development of symmetry ("How"?)

How does the symmetry of the living state develop? With the possible exception of studies of symmetry-reversal in the context of regeneration, this question has not attracted as much interest as the more general question of how biological *patterns* develop. Many symmetric forms can be traced back to the symmetry of the unfertilised egg, and the egg is a highly differentiated product of the mother. This illustrates an important point. Whether we have in mind an egg, embryo or adult, the conventional picture of a living organism as a static configuration is misleading. We are all life cycles – four-dimensional entities, so to speak ("The hen is only an egg's way of making another egg"). Sometimes, when a cylindrically symmetrical egg develops into a bilaterally symmetric adult, one finds that the point where the sperm fertilises the egg, in combination with the cylindrical axis, defines the plane of bilateral symmetry. An external appearance of bilateral symmetry can mask internal handedness (e.g., the human heart is almost always on the left side of the body).

Theories have been put forward, partly based on observations and partly involving testable hypotheses, to try and link internal asymmetry to a polarised spatial distribution of regulatory molecules that is influenced by diffusion, vectorial transport and fluid flow. Two aspects of asymmetry are peculiar to human beings. Both impinge on behaviour, more obviously on motor functions. The first involves handedness in the conventional sense of one hand being preferred over another for performing a whole set of activities. The other, cerebral lateralisation refers to the fact that either the left or the right side of the brain is specialised in respect of certain tasks (e.g. the left side for language). As far as we know, these are independent asymmetries with a weak degree of concordance. On the whole the left hemisphere remains dominant for language in left-handed people, but a higher proportion of left-handers than right-handers have the right-hemisphere dominant for language.

Discussions of the ontogeny of biological symmetry (i.e., how symmetry comes about during development) often lean on a class of models for morphogenesis (the development of form) that the mathematician Alan Turing put forward in 1952 (see http://phylogenous.wordpress.com/2010/12/01/alan-turings-reaction-diffusion-model-simplification-of-the-complex/). Turing's scheme is popular because it is in a sense generic. It depends only on the most general assumptions: the existence of two chemical species that can diffuse within a tissue and possess, respectively, feedback properties implied by the terms short-range activation and long-range inhibition. Crucially, the scheme does not invoke pre-existing inhomogeneity or anisotropy, though cells and tissues may have both (it is often the case that even when a minimal physical model seems to suffice to explain a biological phenomenon, the actual situation contains features – outcomes of evolution – that ensure the reliability of the outcome).

Turing models, and more generally 'reaction-diffusion' models, apply within a specified range of values taken by the parameters of the problem (tissue size and shape, reaction rates, feedback strengths, diffusion coefficients). They show that within this range, a spatially uniform and time-invariant distribution of chemicals becomes unstable with respect to perturbations that are intrinsic to the situation. The reason behind the instability is the random nature of diffusion and chemical reactions. The outcomes can be many. Spatially, one can get a pattern of chemicals that is non-uniform and polarised (with handedness), symmetric or periodic (stripes, spots; see http://www.wired.com/wiredscience/2011/02/turing-patterns/). The temporal outcome can be one of no change (i.e., a stable steady state) or one that exhibits periodic variations. In combination, travelling waves can result. A large number of phenomena have been explained on the basis of Turing-type models. The closest link-up with observational data comes from models for the origin of periodic stripes on the skin of certain fish and for the symmetrical positioning of the plane of cell division in bacteria.

The handedness of the shell in some snails depends on the genetic constitution of the snail's mother. But in most cases the genetic basis of biological symmetry is unknown. In any case, the number of genes and gene products involved may be so large as to make it uninteresting to theorise about a genetic basis. Situs inversus, an extremely rare condition in human beings that results in a mirror-image transposition of the heart and other internal organs, may be an exception to the rule (such people are symptom-less and tend to remain undetected until their condition is picked up during routine medical examination; see http://www.rndsystems.com/mini_review_detail_objectname_MR03_TGF-betaLigands.aspx). The reason for thinking so is that there is a mutation in the mouse that gives rise to the same outcome. At the same time, it is also the case that alterations in genetic states can affect symmetries in striking ways.

An interesting generalisation is that when a mutation affects symmetry, it does so by raising the symmetry, not lowering it. Normal (wild-type) fruit fly larvae are segmented along their body length. Superficially, many of the segments look the same. But they display differences on closer examination; these differences are lessened in mutant larvae. Along similar lines, the normal fruit fly has one pair of wings attached to a thoracic segment and a pair of balancers attached to the next posterior segment. A mutant can be four-winged, with the balancers having been converted to a second pair of wings (see http://www.mun.ca/biology/scarr/Bithorax_Drosophila.html).

An example of temporal symmetry is the following. A particular round of cell division in the normal nematode worm is asymmetric and the two daughters are both different from the mother. In a mutant worm, on the other hand, the same cell division can yield one differentiated daughter. The other daughter resembles the mother and reiterates her pattern of cell division. Such instances are significant when viewed from an evolutionary viewpoint. Modern flies (dipterans) are believed to have evolved from four-winged insects, and the insects in turn are believed to have derived from animals like millipedes – or according to some views crustaceans – which show less dramatic differences between their body segments. Thus in these cases the mutants just described resemble atavisms; that is, they seem to hearken back to an ancestral state. Because we believe that evolution has involved the coming into play of new genetic capabilities, the mutants hint that at least in respect of some features, the evolution of new forms has been accompanied by new genetic functions that foster symmetry-breaking transformations.

Evolution of symmetry ("Why?")

Why is biological symmetry an almost universal feature of life? As indicated earlier, we can look for answers based on physics and chemistry alone. On the other hand, experience shows that before searching for alternatives, one should speculate to what extent natural selection might provide an answer. That in turn forces one to think of the possible advantages of symmetry. Here, advantage has to be understood in terms of a heritable component of reproductive fitness.

One has to think of both how symmetry may have originated during evolution and what might be responsible for its persistence. If cellular life began in the ocean, as is generally believed, the first question may have an easy but non-evolutionary answer. Minimum energy considerations involving surface tension and fluid pressure would have led the very oldest cells to adopt spherical shapes.

The real issue concerns whatever might have lead to deviations from spherical symmetry. Here we turn to the evolution of sexual dimorphism, which led to the appearance of two distinguishable sexes (see http://en.wikipedia.org/wiki/Sexual_dimorphism). One of them, the female, produced a large egg cell and endowed it with nutrients. If the nutrients were made of macromolecules, as in yolk, they would separate from the rest of the egg on account of their differential buoyancy. A separation would also result if the mother injected them into a restricted region within the egg. In either event, the egg would acquire cylindrical symmetry. There is a plausible argument for the origin of bilateral symmetry too. Given that it is of advantage to cover the distance between two points in the shortest time, moving in a straight line would be favoured over moving in an arc. A bilaterally symmetric arrangement of the organs of motility would be economical. By providing equal motive forces on two sides of the body, it would allow for rectilinear motion. At the same time, it would not compromise on the importance of receiving sensory inputs from as wide an angle as possible. The normal movement pattern of amoeboid cells is random, but they can home in towards targets (e.g. food). They and many creatures that make use of flotation (e.g. jellyfish) have retained cylindrical symmetry.

We have seen that there can be asymmetry at the level of the individual but not at the level of the population, and asymmetry at the level of the population (i.e., a predominant handedness). When found along with population-level symmetry, individual-level asymmetry has been explained on different grounds. In some flatfish, a free-swimming adult drops down, as it were, to the ocean floor and adopts a largely sessile life. Whether it chooses to rest on its left or right side is a matter of accident. The (new) lower eye and mouth move around to the top after the choice has been made (see http://biochemicalsoul.com/2009/02/adaptation-of-the-week-flatfish-recapitulation/). In some snails, the helicity of shell coiling in males and females has to be the same if mating is to take place successfully.

On the other hand, there could be situations in which the requirements of mating demand complementary geometries, and therefore oppositely handed individuals. In the case of lobsters and fiddler crabs one of the two anterior-most claws – it does not matter which one – grows to an enormous size and serves as a signal for sexual attraction (see http://etc.usf.edu/clipart/8200/8250/fiddler_crab_8250.htm/). When the large claw is also used as an organ of combat, it may help individuals to develop it on the same side of the body as the prevailing majority; in that case one would get entire populations of left- or right-handed individuals.

Molecular biology has provided an unexpected insight into the evolution of asymmetry in certain fish lineages. After examining an evolutionary tree based on DNA sequences, it turns out that in certain lineages where the present condition is one of a predominant handedness; the ancestral situation was composed of individual asymmetry but population-level symmetry (see Figure 1 in http://www.biology.ualberta.ca/palmer/pubs/96PNAS/96PNAS.htm). A plausible inference would be that the original situation derived not so much on genetic specification as on a scheme for spontaneous symmetry breaking that was indifferent to handedness (e.g., a Turing mechanism), and that was followed selection for a particular handedness.

The most intriguing, if also most controversial, evolutionary speculations involving symmetry revolve around fluctuating asymmetry (*FA*). As we have seen, the term refers to the commonly seen deviation from perfect bilateral symmetry in individuals (e.g. the human face). Typically, the deviations are quantified by an index defined as the absolute value of the difference in some measurable trait for example, the number of hairs or the area of a coloured patch – between the two sides of the body divided by its average on the two sides: in symbols, $FA = 2|L - R|/(L + R)$. Fluctuating asymmetry is ascribed to the fact that a host of chance factors, all unavoidable, impinge on development. (This is sometimes referred to as developmental noise.) As a result the precision with which any feature develops is invariably less than 100%. Turning the reasoning the other way around, one can argue that individuals who show relatively low levels of fluctuating asymmetry must have incorporated a high level of quality control. Their symmetry is a signal of fitness, because they are showing that they are capable of investing the resources required (eventually, presumably in terms of energy) to fine-tune developmental processes. The ability to detect and respond to the signal would be advantageous to potential mates as well as potential competitors. Some experiments show that such is indeed the case; females prefer males of higher symmetry. A similar tendency is said to exist – in both directions – in human beings too, and has been seized upon by psychologists with an evolutionary bent of mind to propose explanations that have excited much debate but little agreement. A. J. S. Klar has put forward an ingenious hypothesis that links handedness, the coiling of scalp hair whorls, and genetics (see http://www.dana.org/news/cerebrum/detail.aspx?id=656). Long ago, A. H. Sturtevant showed that curious patterns shown by the inheritance of handedness of shell coiling in snails could be explained as being due to the action of different states of the same gene – with the twist, so to say, that the effect of the gene was felt after a time-delay of one generation.

Yellow dung flies of the species *Scatophaga starcoraria* tend to congregate around fresh cattle droppings, males outnumbering females 4 to 1. The aim of the game – for a male – is to find a female that is ready to lay eggs and to mate with her, thereby passing down his own genes to posterity. Females, being in a minority, can afford to be choosy. Evolution will favour those females that can pick the best males to mate with. On what basis do they decide who the best males are? Researchers from the University of Liverpool (presumably with strong noses) made careful measurements of the vital statistics of three classes of males: those on a dropping that had paired with a female, those on a dropping but unpaired and solitary males from the surrounding grass. What they found was that the males belonging to the first, successful, group had markedly similar left and right halves, but the males in the other two groups were quite asymmetrical. This was true of the lengths of two structures that were monitored, the wing and the tibia (a segment of the foreleg).

What do these findings mean? Females obviously are choosing the most symmetrical males, but why? One possibility is that the more symmetrical a male, the larger it is: one can plausibly argue that a large male has "better" genes than a small one. But measurements did not bear this out: there was no correlation between degree of symmetry and body size. In fact, the largest males of the lot were those that were on a dropping but unsuccessful in pairing.

Two answers have been offered as solutions to the puzzle. One, suggested by the Liverpool scientists, is that symmetrical males are better at executing the complex manoeuvres associated with flying that are so essential for a fly successfully to complete for access to a female. For example, what aeronautical engineers call the wing load turns out to be smallest for the successful males, slightly higher for unsuccessful males on the dropping and highest of all for males that are just hanging around in the neighbourhood. The differences appear to be significant, as much as 25 per cent of the average.

A different argument has been advanced by the Israeli evolutionary biologist A. Zahavi, and I leave you to decide which one is more plausible, or whether both are equally plausible. Zahavi thinks that form and visual appearance act as powerful signals in themselves. They convey to an observer an accurate indication of one's real worth, which in the ultimate analysis means one's genetic quality. It is important for signals to be intrinsically reliable, because otherwise it becomes easy to manipulate others by sending misleading signals. In other words the signals that evolve are such that (a) only honest signallers can afford to develop them; (b) cheating is a prohibitively expensive option (see http://www.ias.ac.in/resonance/May2010/p434-440.pdf). When Malleshwari lifted her weights we knew for sure that she had powerful muscles, just as we know for sure that a five-star hotel wedding means that someone is very rich. So a high degree of symmetry implies that the plant or animal is built with precision tools that have low error tolerances in terms of their genetic makeup – in short, with good genes.

Summary

Few things are more pleasing to the eye than symmetry. Whether it is a flower, a bird or another human being, almost nothing excites our attention and quickens our interest as much as a symmetrical appearance. The opposite is true as well. Blatant lack of symmetry can be jarring, which is why Picasso's paintings are an acquired taste.

Usually we take 'symmetry' to mean that the left and right halves of something look the same, but in scientific use the word has a broader connotation. The technical definition of symmetry always relates the term to a specific operation. For example, a perfect sphere looks the same however you turn it around; it is said to have rotational symmetry. Railway lines look much the same as you travel along them, and they are said to have translational (displacement-related) symmetry.

As the last example suggests, symmetries in real life are approximate: the line ends somewhere, or branches out, or whatever; you start noticing differences. Living creatures exhibit all sorts of symmetries and these too are approximate, as we well know. Our left and right halves are not exactly identical, each petal on a flower does not match the others perfectly and the segments of an earthworm can be told apart (with some effort).

There has been a long but intermittent interest among biologists in symmetry. Developmental biologists have been engaged on the question of what brings about a symmetric form in an animal or plant; evolutionary biologists want to know why symmetry should be there at all. Part of the response to the first problem is that the question is wrongly posed. The fact of the matter is that various symmetries do not so much appear as disappear during the transformation of egg to adult (which therefore is in a sense a regressive, not progressive change). As for the evolutionary reason, it appears that symmetry is there because it is good for us... or at any rate, good for swallows, dung flies and many other plants and animals. What we refer to as the pleasing quality of symmetrical objects, and as our aesthetic preference for beauty, may have distant evolutionary antecedents.

Acknowledgements

I thank Prof. I. B. S. Passi for inviting me to the symposium at which these ideas were presented. Portions of the text have been taken from an article of mine that appeared in the Economic Times of 29 April 1995 as one among a series of essays titled 'Date With Darwin'.

Further Reading

(a) *General*

"Specific Symmetry of Living Systems"; V. A. Gusev (Chapter 9 of "Progress in Biological Chirality", G. Palyi, C. Zucchi and L. Caglioti, Eds., Elsevier, 2004).

"Eclectic Reflections on Biological Asymmetry"; A. R. Palmer (http://www.biology.ualberta.ca/palmer/asym/asymmetry.htm).

"Organized mayhem in Bilateria Baupläne: symmetry and animal complexity"; D. Rasskin-Gutman, Coloquios de Paleontología, Vol. Ext. 1, 559-567 (2003). (http://revistas.ucm.es/geo/11321660/articulos/COPA0303220559A.PDF)

"Symmetry"; H. Weyl, Princeton University Press (1983).

"Symmetry in Biology"; http://en.wikipedia.org/wiki/Symmetry_in_biology.

(b) *'Turing' and other self-organised patterns*

"Biochemical Oscillations and Cellular Rhythms"; A. Goldbeter, Cambridge University Press (1996).

"Reaction-Diffusion Model as a Framework for Understanding Biological Pattern Formation"; S. Kondo and T. Miura, Science 329 (5999), 1616–1620, 2010.

"Traveling stripes on the skin of a mutant mouse"; N. Suzuki, M. Hirata and S. Kondo, Proc. Natnl. Acad. Sci. USA 100 (17): 9680-9685 (2003).

"Adenosine 3', 5'-monophosphate waves in *Dictyostelium discoideum*: a demonstration by isotope dilution—fluorography"; K. J. Tomchik and P. N. Devreotes, Science 212 (4493): 443–446 (1981).

"The Chemical Basis of Morphogenesis"; A. Turing, Phil. Trans. R. Soc. London *B* 237, pp. 37–72 (1952).

"Alan Turing and *The Chemical Basis of Morphogenesis*"; V. Nanjundiah, in "Morphogenesis and Pattern Formation in Biological Systems", pp. 33–44, T. Sekimura, ed., Springer-Verlag, Tokyo (2003).

(c) *Symmetry and evolution*

"Symmetries throughout organic evolution"; A. García-Bellido, Proc. Natnl. Acad. Sci. USA 93 (25): 14229–14232, 1996.

"Human Handedness and Scalp Hair-Whorl Direction Develop From a Common Genetic Mechanism"; A. J. S. Klar, Genetics 165, 269–276 (2003).

"The evolutionary psychology of facial beauty"; G. Rhodes, Annual Review of Psychology 57, 199–226, 2006.

"Courtship and genetic quality: asymmetric males show their best side"; M. R. Gross, H. Y. Suk and C. T. Robertson, Proc. R. Soc. *B* 274, 2115–2122 (2007).

"The Genetics and Evolution of Fluctuating Asymmetry"; L. J. Leamy and C. P. Klingenberg, Annual Review of Ecology, Evolution, and Systematics, 36, pp. 1–21 (2005).

"Preferences for symmetry in faces change across the menstrual cycle"; A. C. Little, B. C. Jones, D. M. Burt and D. I. Perrett, Biological Psychology 76: 209–216, 2007.

"Patterns of fluctuating asymmetry in sexual ornaments predict female choice"; A. P. Møller, J. evol. Biol. 6: 481–491 (1993) (also available in http://onlinelibrary.wiley.com/doi/10.1046/j.1420-9101.1993.6040481.x/pdf).

"Bilateral symmetry and sexual selection: A meta-analysis"; A. P. Møller and R. Thornhill, American Naturalist 151, 174–192, 1998.

"Evolutionary developmental genetics of floral symmetry: the revealing power of Linnaeus' monstrous flower"; G. Theissen, BioEssays 22: 209–213 (2000).

"Lateralization of brain function"; http://en.wikipedia.org/wiki/Lateralization_of_brain_ function.

"The handicap principle: a missing piece of Darwin's puzzle"; A. Zahavi and A. Zahavi, Oxford University Press, New York, NY (1997).

Appendix

Reaction-Diffusion ('Turing') models

This is a brief sketch of the scheme formulated by Turing. Details can be found in the original article and in the references listed below. Consider a homogeneous tissue; for simplicity, let us think of it as a line of cells of length L. The chemical species of interest to us has a concentration, i.e. number of molecules per unit length, denoted by C. Its concentration leads – by means that do not concern us here – to a certain state of differentiation in the cell. For example, the chemical may stimulate a pathway of pigment synthesis such that a pigment is deposited whenever its concentration exceeds a threshold. If the stable distribution of the chemical resembles a series of waves of sufficient amplitude, the pigment pattern will resemble the stripes of a zebra. A chemical substance with this property of evoking a spatially patterned response is known as a *morphogen*.

C is made and broken down by various reactions whose net effect is described by a rate-function $R(C)$; note that R is assumed to be independent of both where the chemical is made and when it is made. Also, the chemical diffuses freely through the tissue with a diffusion coefficient D – this is the simplest form of transport that one can think of. At the boundaries of the tissue there is a steady leak of the chemical (either into the tissue or away from it) at a rate that is proportional to the concentration; the boundaries are assumed to be equivalent. The system can be modelled with the help of the standard form of a 'reaction-diffusion' equation in 1-dimension with symmetrical boundary conditions. C can change over time in any cell on account of one or both of two reasons: diffusion to or away from the cell and synthesis or breakdown. For small concentration gradients Fick's law of diffusion holds, and the diffusive flux J is given by $-D.\,\partial C/\partial x$. By the equation of continuity, $\partial C/\partial t = -\partial J/\partial x + R(C)$. This gives us

$$\partial C/\partial t = \partial/\partial x(D\partial C/\partial x) + R(C), \quad 0 < x < L;$$

$$\partial C/\partial x = hC \text{ at } x = 0 \quad \text{and} \quad x = L \text{ where } h \text{ is some constant.}$$

If more than one chemical species is under consideration, C is the vector of chemical concentrations and R that of reactions; D becomes a constant matrix of diffusion coefficients that includes both self-diffusion terms D_{ii} (with the flux of a chemical species C_i depending on its own spatial concentration gradient) and cross-diffusion terms D_{ij} (in which the flux of a chemical species C_i is driven by the spatial concentration gradient of another species C_j). There are two ways of tackling the problem. The first, due to Turing (1952), examines instabilities (if any) that occur in the system when it is perturbed about a spatially uniform steady state. The wavelength of the most rapidly growing instability is assumed to set the scale of pattern, at least for short times. The second approach is directly to look for stable solutions of the time-independent equations. We will follow the first approach.

The basic results are as follows. It turns out that if one makes the reasonable assumption that the tissue is self-contained, i.e. if the boundary conditions are of the no-flow type ($\partial C/\partial x = 0$ at $x = 0$ and $x = L$), and if one is looking for solutions that vary in space, there must be at least two chemical species. A one-chemical system cannot support a stable, spatially non-uniform concentration distribution (Casten and Holland 1978, Fife 1979). With a two-component system, no-flow boundary conditions and no cross-diffusion, the equations can be written as

$$\partial C_1/\partial t = D_1\partial^2 C_1/\partial x^2 + R_1(C_1, C_2),$$

$$\partial C_2/\partial t = D_2\partial^2 C_2/\partial x^2 + R_2(C_1, C_2)$$

$$\text{for} \quad 0 < x < L, \quad \text{and}$$

$$\partial C_1/\partial x = \partial C_2/\partial x = 0$$
$$\text{at } x = 0, L$$

Now one assumes that the two rate functions R_1, R_2 can vanish simultaneously at some values of $C1$ and $C2$. In other words, the equations admit spatially uniform solutions $C_1 = C_1^*$, $C_2 = C_2^*$, and the system initially is in the state so defined:

$$R_1(C_1, C_2) = R_1(C_1, C_2) = 0 \text{ at } C_1 = C_1^*, \quad C_2 = C_2^*$$

If these solutions (there may be more than one) are also stable, the system will remain in a uniform steady-state for all time. But if the solutions are unstable, one may hope that the instability leads to an interesting spatially structured pattern. As a means of looking for instability we put

$$C_1 = C_1^* + \delta C_1, \quad C_2 = C_2^* + \delta C_2,$$

with $\delta C_1(x, t)$ and $\delta C_2(x, t)$ standing for spontaneously occurring fluctuations around C_1^* and C_2^*. The fluctuations could be thermal in origin or could be due to the inherently probabilistic nature of diffusion or the chemical reactions (i.e. could come in via R_1 and R_2). $\delta C_1 \ll C_1$ and $\delta C_2 \ll C_2$ in absolute value, so that to first order one can write

$$R_1(C_1, C_2) = R_1(C_1^*, C_2^*) + (\partial R_1/\partial C_1) \cdot \delta C_1 + (\partial R_1/\partial C_2) \cdot \delta C_2,$$
$$R_2(C_1, C_2) = R_2(C_1^*, C_2^*) + (\partial R_2/\partial C_1) \cdot \delta C_1 + (\partial R_2/\partial C_2) \cdot \delta C_2$$

where all four partial derivatives are to be evaluated at $C_1 = C_1^*$, $C_2 = C_2^*$. We know that $R_1(C_1^*, C_2^*) = R_2(C_1^*, C_2^*) = 0$; using this fact and restricting ourselves to first-order terms, we get a pair of linear partial differential equations in δC_1 and δC_2. They are

$$\partial \delta C_1/\partial t = D_1 \partial^2 \delta C_1/\partial x^2 + (\partial R_1/\partial C_1) \cdot \delta C_1 + (\partial R_1/\partial C_2) \cdot \delta C_2,$$
$$\partial \delta C_2/\partial t = D_2 \partial^2 \delta C_2/\partial x^2 + (\partial R_2/\partial C_1) \cdot \delta C_1 + (\partial R_2/\partial C_2) \cdot \delta C_2$$

For the sake of simplicity, let us put

$$(\partial R_1/\partial C_1)_{(C1^*,C2^*)} = \alpha, \quad (\partial R_1/\partial C_2)_{(C1^*,C2^*)} = ß,$$
$$(\partial R_2/\partial C_1)_{(C1^*,C2^*)} = \gamma, \quad (\partial R_2/\partial C_2)_{(C1^*,C2^*)} = \delta$$

and look for a general solution in terms of Fourier expansions of δC_1 and δC_2:

$$\delta C_1(x, t) = \Sigma C_{1n} \exp(ik_n x + \omega_n t), \quad \delta C_2(x, t) = \Sigma C_{2n} \exp(ik_n x + \omega_n t),$$

where the index n represents the nth Fourier mode, $2\pi/k_n$ is the wavelength of that mode and ω_n is its (complex) rate of growth. Re $\omega_n = 0$ implies that a fluctuation of that wavelength neither grows nor decays, Re $\omega_n < 0$ implies that it decays (the steady-state is stable with respect to such a perturbation) and Re $\omega_n > 0$ implies that it grows (i.e., the steady-state is unstable with respect to such a perturbation). After substitution, we see that the modes separate out because the equations for δC_1 and δC_2 are linear. That enables us to write

$$\omega_n C_{1n} = -D_1 \cdot k_n^2 \cdot C_{1n} + \alpha \cdot C_{1n} + \beta \cdot C_{2n},$$

$$\omega_n C_{2n} = -D_2 \cdot k_n^2 \cdot C_{2n} + \gamma \cdot C_{1n} + \delta \cdot C_{2n}.$$

These are a pair of linear homogeneous equations in C_{1n} and C_{2n}. For a non-trivial solution to exist, ω_n has to satisfy the quadratic equation

$$(\omega_n + D_1 \cdot k_n^2 - \alpha) \cdot (-\omega_n - D_2 \cdot k_n^2 + \delta) + \beta \cdot \gamma = 0$$

If the linearised (i.e. perturbed) equations are to have an unstable non-uniform mode that grows steadily, ω_n must be real and positive for some non-zero value of n. This means that a whole set of requirements must be satisfied. Going back to the original symbols, these are

$$D_1 \neq D_2, (\partial R_1/\partial C_2) \cdot (\partial R_2/\partial C_1) < 0, (\partial R_1/\partial C_1) \cdot (\partial R_2/\partial C_2) < 0$$
$$\text{and} \quad \partial R_1/\partial C_1 + \partial R_2/\partial C_2 < 0$$

(Turing, 1952, Maynard Smith, 1968). As a parenthetical remark, it is interesting that the conditions are the same if R_1 and R_2 are linear functions and we look for a *stable* time-independent solution (Casten and Holland, 1978). Our analysis tells us whether a particular mode is unstable; it says nothing about what happens after a mode has grown so much that the first-order approximation is no longer valid. Except for those rare cases in which the non-linear reaction-diffusion equations can be solved exactly, only computation can reveal the final spatial pattern. In many cases the computed solutions do display the features of regularity and symmetry that are seen in plant and animal patterns. Note that if Re $\omega_n > 0$ and Im $\omega_n \neq 0$, an oscillatory instability is possible; one can get 'travelling wave' patterns. If Re $\omega_n > 0$ for more than one value of n (as is likely to be the case), the mode that dominates initially is the one that grows fastest, i.e. the one for which Re ω_n has the largest positive value.

Gierer and Meinhardt (1972) have made a detailed study of this sort of scheme (with additions involving inhomogenous source terms), which they call 'short-range activator-long-range inhibitor' models. The reason for the names becomes clear on considering the specific combination of conditions given by

$$D_1 < D_2, \quad \partial R_1/\partial C_1 > 0, \quad \partial R_2/\partial C_2 < 0, \quad \partial R_1/\partial C_2 < 0, \quad \partial R_2/\partial C_1 > 0.$$

To take them one by one, $\partial R_1/\partial C_1 > 0$ means that a slight increase in C_1 leads to a further increase in C_1(R_1 being the net rate of production of C_1); $\partial R_2/\partial C_1 > 0$ means that a slight increase in C_1 leads to an increase in C_2. Therefore C_1 'activates' the production of both itself and C_2. Similarly, $\partial R_1/\partial C_2 < 0$ and $\partial R_2/\partial C_2 < 0$ mean that a slight increase in C_2leads to a decrease in the concentrations of C_1 and C_2; thus C_2 'inhibits' the production of both itself and C_1. Also, $D_1 < D_2$ implies that because of its smaller coefficient of diffusion the effective range of C_1 is less than that of C_2.

Thus the description 'short-range activation' combined with 'long-range inhibition' is appropriate.

Do such systems exhibit scale-variance? No, except approximately – that is to say, for a restricted range of lengths within which the same unstable mode of growth (i.e. same value of n) dominates. In other words, the range of lengths over which a perturbation corresponding to one of the normal modes of the system is unstable, does not overlap with the range of "unstable lengths" for any other mode (see Lacalli and Harrison, 1978, for details). Because the original equations have diffusion coefficients combined with the second derivative of x, one would need D_1 and D_2 to be proportional to L^2 for perfect scale invariance. Limited scale-invariance can also be demonstrated in the schemes proposed by Edelstein (1972) and Gierer and Meinhardt (1972), though these involve non-uniform initial distributions of morphogen, inhomogeneities in the system, or both. Cohen (1972) has obtained limited scale-invariance by using the assumption that morphogen transport depends on both passive diffusion and – for low concentration gradients – polarized active transport. A case of *exact* scale-invariance in a reaction-diffusion scheme has been proposed by Othmer and Pate (1980). In addition to the morphogens, they postulate the existence of a controlling chemical which is synthesized at a constant rate everywhere and influences the diffusion of the morphogens. If the diffusion constants of the morphogens vanish in the absence of the controlling species, and if the boundaries of the system are infinitely permeable to the controlling chemical, exact scale invariance results.

References

[1] R. G. Casten and C. J. Holland, Instability results for reaction-diffusion equations with Neumann boundary conditions, *J. Diff. Equns.*, **27** (1978) 266–273.

[2] M. H. Cohen, Some mathematical questions in biology (J. D. Cowan, ed., American Mathematical Society, Providence, Rhode Island, USA) (1972).

[3] B. B. Edelstein, The dynamics of cellular differentiation and associated pattern formation, *J. theor. Biol.*, **37** (1972) 221–243.

[4] P. C. Fife, Lecture notes in biomathematics (American Mathematical Society; Providence, Rhode Island, USA) (1979).

[5] A. Gierer and H. Meinhardt, A theory of biological pattern formation, *Kybernetik*, **12** (1972) 30–39.

[6] T. C. Lacalli and L. G. Harrison, The regulatory capacity of Turing's model for morphogenesis, with application to slime molds, *J. theor. Biol.*, **70** (1978) 273–295.

[7] B. L. Lokeshwar and V. Nanjundiah, Biological Patterns and the Problem of Regulation (in The Living State, Vol. 1; R. K. Mishra, ed., Wiley (1984).

[8] J. Maynard Smith, Mathematical Problems in Biology, Cambridge University Press (1968).

[9] H. G. Othmer and E. Pate, Scale-invariance in reaction-diffusion models of spatial pattern formation, *Proc. Nat. Acad. Sci. USA.*, **77** (1980) 4180–4184.

Symmetry: A Multi-Disciplinary Perspective No. 16, 2011, pp. 49–62.

Encoding and Transcending Symmetry: The Dynamics of Space Conception and Perception in Architecture

Yatin Pandya

Footprints E.A.R.T.H. (Environment Architecture Research Technology Housing), Milan Bunglow, Sargam flat lane, Ishvarbhuvan to stadium road, Navjivan post, Ahmedabad 380 014, India
e-mail: pandyatin@hotmail.com; pandyatin@gmail.com; footprintsearth@gmail.com

Abstract. The paper is an exploration of the phenomena of Symmetry as observed, interpreted and applied in architecture. It observes that symmetry is well integrated in nature and nature has been the source of inspiration for architectural resolutions. Architecture has used diverse forms, attributes and principles of symmetry in space organization to derive desired impact for its perceiver; such as formality, balance, equity, dignity, simplicity, directionality and so on. The paper argues that architecture is perceived not from a static point of view but from changing positions while in movement. This kinesthetic perception makes it a complex process of communication between architect as encoder of messages and perceiver as the decoder of the same. Paper interprets that in good, experientially engaging architecture, principles of symmetry are used in plan organization while conceiving the space but it transcends the staticness and predictability of symmetry during its dynamic perception through 'kinesthetics', offering the ever changing views and invoking sense of curiosity.

Keywords. Symmetry, architecture, space perception, kinesthetics, timeless architecture, conception vs. perception.

Symmetry: Phenomena

Symmetry generally conveys two primary meanings. The first meaning is a precise and well-defined concept of balance or "patterned self-similarity" that can be demonstrated or proven according to the rules of a formal system: by geometry, through physics or otherwise. While more generic and commonplace meaning of the term is an imprecise sense of harmonious or aesthetically pleasing proportionality and balance such that it reflects beauty or perfection. [1]

There are different kinds of symmetries:

A) Bilateral Symmetry – Mirroring along the axis as reflection
B) Radial symmetry – Rotation in radiating lines around a point
C) Similar Symmetry – Not precisely identical but similar in essence
D) Fractal symmetry – Replication in radiating as well as growing order
E) Notional Symmetry – Not graphically same but similar in spirit

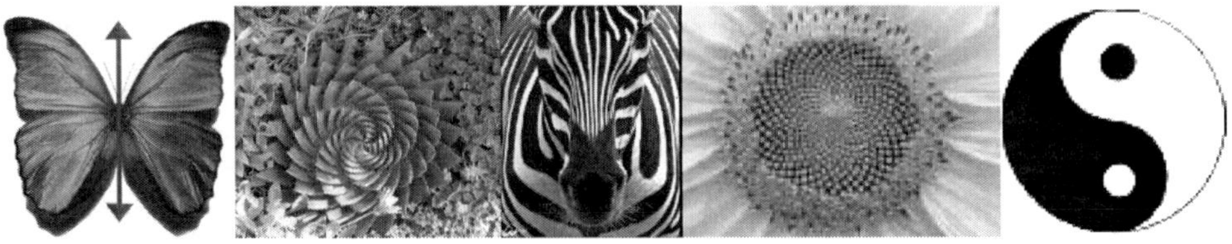

Bilateral symmetry | Radial symmetry | Similar symmetry | Fractal symmetry | National symmetry

Symmetry in nature

Symmetry is one of the most common place, inherent and manifest principles of organization abundantly prevalent in nature. May it be Flora in terms of seeds, flowers, fruits or even trees, may it be Fauna in terms of insects or the entire animal kingdom, even the human beings from the micro cell to face to the entire body; or may they be the molecular structure of the elements as well as physical elements such as crystals, corals and conches, all pervade through the principles of symmetry. Even the diurnal cycles of time or annual cycles of seasons imply replicative temporal symmetry.

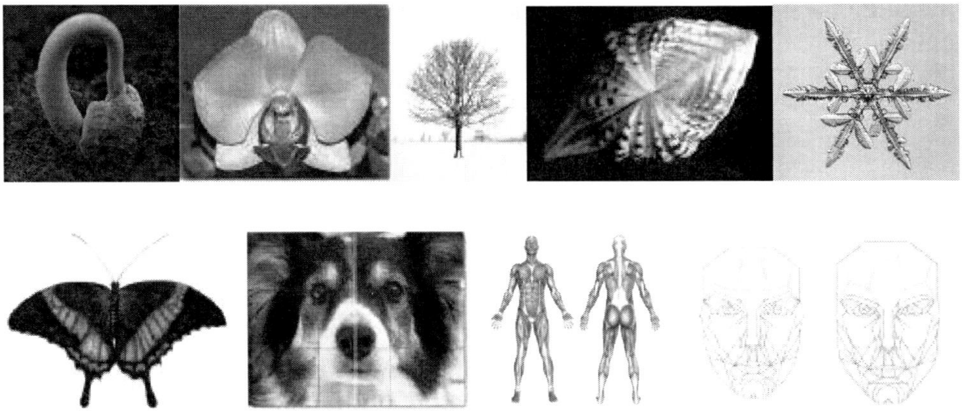

Symmetry in nature

Having been born of nature, as an aspect of nature, we perceive symmetry more naturally. In that sense, symmetry is the most primordial phenomena of organization.

Attributes of symmetry

The phenomenon of symmetry inherently imbues certain attributes:

a) Equity and Balance
   As symmetrical organization represents replication of one side on to the other, it naturally implies the balance. Being balanced and equally distributed along the axis it represents equality.

b) Directionality and Continuity
Having been organized along the axis, the axis signifies the direction. The line of axis forms the direction of its organization. As the axis is unending it offers the sense of continuity as well as infinity.

c) Formality and Discipline
As the principle of Symmetry inherently follows the principles of similarity along the axis, it leaves very little room or chance for manipulation, thus rendering it formal. The formal and defined way of organization makes it disciplined as it adheres to set structure, living no room for deviation.

d) Dignity and Grandeur
Aspect of formality and discipline lead to dignity and implied Grandeur

e) Simplicity and Comprehensibility
   As the organizational principle groups two sides equally and opposite one can take one half for granted as per the other and this makes the comprehension easy and simple.

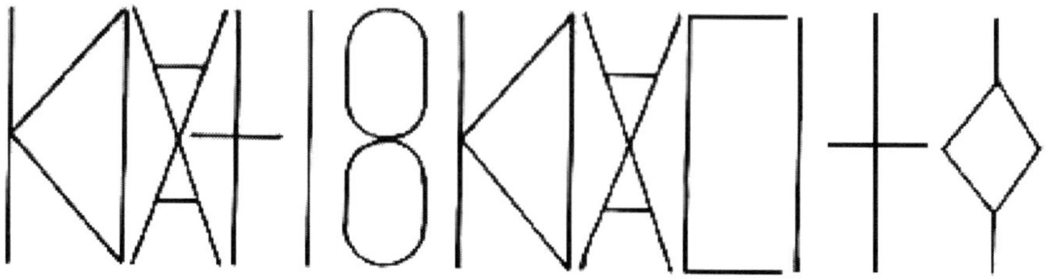

f) Opposition
   While, the equal distribution along the axis suggests balance, its mirroring implies the opposition.

It may not have been a matter of mere coincidence that most religious symbols of the world or the religious shrines have organized around symmetry, exploiting these attributes of equity, balance, dignity, infinity and continuity.

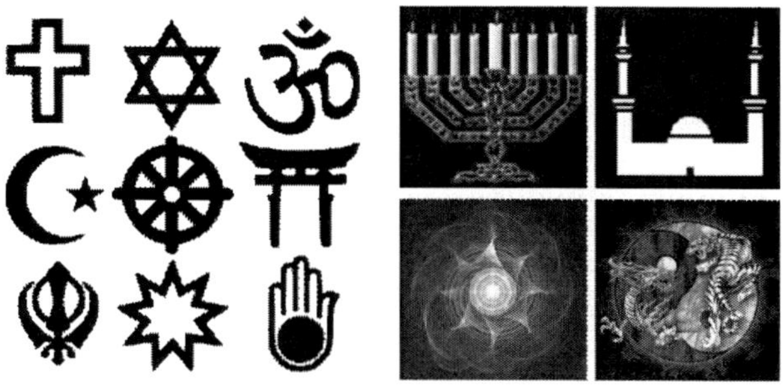

Symmetry as applied in architecture

No wonder, these attributes of symmetry inspire architecture immensely, as architectural spaces can potentially nourish emotionally and spiritually. These aspects, attributes and properties of symmetry have been effectively applied in architecture all along its passage of time and across its geographical expanse. Inspired it may be by

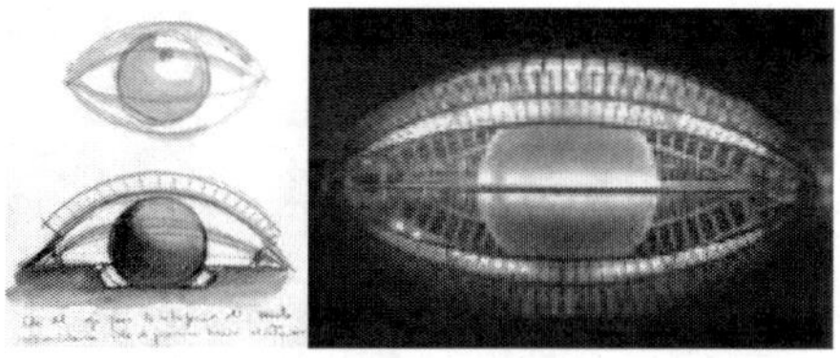

Eye as inspiration

Tree as inspiration

Lotus as inspiration

the radial and fractal symmetry of the form of the flower, or the bilaterally symmetrical organisational structure of the tree, or the symmetrical arrangement of elements in biological organism, architecture has imbibed the principles of nature from its form manifestation to organisational structure to its detailing.

From the very first manmade structure as trabeated construction of the beam spanning over the bilaterally symmetrically placed columns, or the corbelled stone

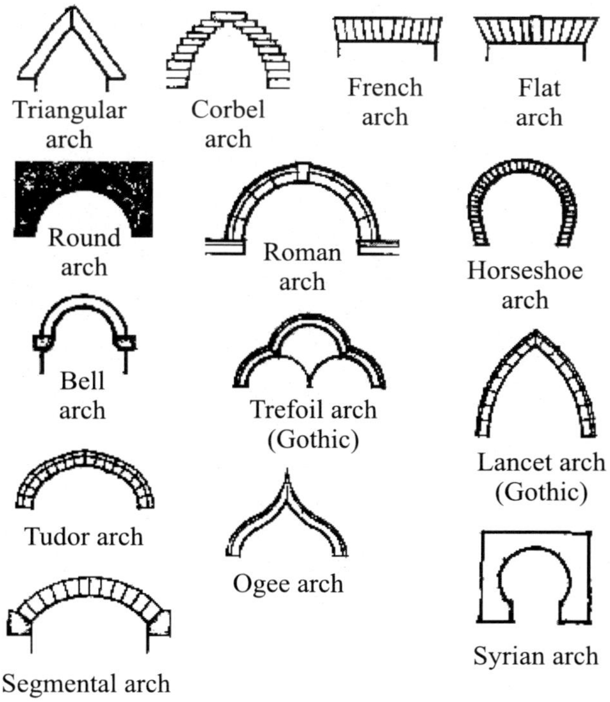

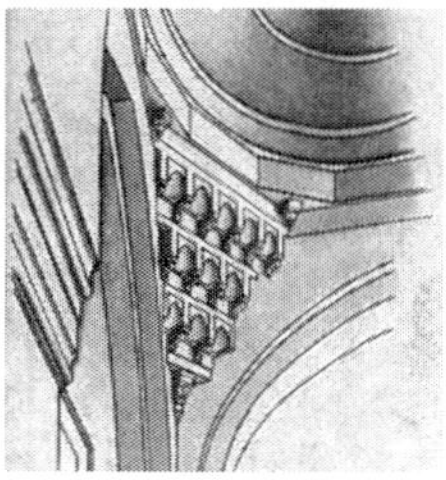

Sqinches

domes with radially symmetrical rotation of squares, to roman arches, vaults and dome rotated along the *Y* axis, to Islamic pointed arches with bilateral symmetry, to Mughal quiches as fractal symmetry and subsequent European segmental and multi point arches as reflective symmetry, all have been exploited properties of symmetry in spanning the space by resolving the structure through its fundamental attributes of balance. These principles have been further evolved and creatively applied to most advance spanning structural systems in contemporary times. Space frames, grid shells, geodesics or the folded plate structures over the sports stadia and congregational arena have ably demonstrated the application of this phenomenon.

Space frame

Grid shell

Geodesic dome

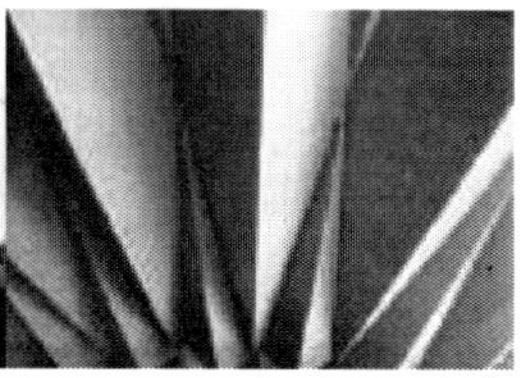
Folded plate structure

Not only in physical support structure but even in evoking desired ambiance as well as associations, the principles and attributes of symmetry are effectively imbibed. Chapels and churches bring out the sense of formality through bilateral symmetry and axiality pointing to the direction of nave and altar. Idea of charbagh and paradise garden in Islamic landscaping as well as architecture of mausoleums to denote equity and centrality use the principles of cardinal/ bi directional symmetry. Lotus form of a building with radiating petals as roof shells for the associational value of the spiritual abode all have integrated aspects of symmetry for the emotional, experiential as well as metaphysical overtones of architectural spaces.

Char baugh garden

The conception versus perception of architecture

Architectural design is a two way communication, where the designer encodes the messages while the perceiver decodes the same. Symmetry, for the ease of its comprehension and simplicity of its resolution, is quite commonly employed in architectural designs. Especially, in the plan diagrams.

Diagram 1
The temple as it visually appears with correction

Diagram 2
The temple as it would appear without correction

Diagram 3
The temple as it is actually built with correction

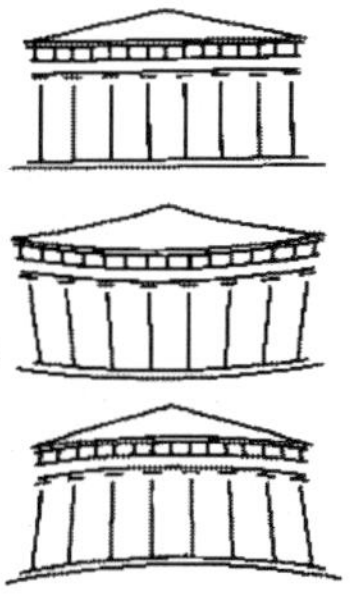

Optical correction (Entasis) in doric temples

Use of geometry in proportioning the building and perspectival visual corrections as applied in early greek architecture

Kinesthetic perception in architecture

What complicates the matter most in architecture is the kinetic nature of space perception. Unlike painting or a theatre, architecture is not perceived from static, defined singular point of view. Architecture is perceived kinetically, while in movement, where point of view, the reference point of visual perception constantly changes with changing position of the perceiver. While the building is static-immobile-the perceiver is constantly moving. He sees the building from distance, he approaches close, he enters the building, moves within and exits . . .. Thus, he sees the same physical construct from different spatial co-ordinates. This phenomenon of kinaesthetic, dynamic perception in movement render architectural spaces into event where experience of the space is a unique chemistry of perceiver relative to his position in a given time and space. It is this aspect of visual perception that adds dynamism to architectural spaces. Symmetrical space organisation in planar constellation can appear asymmetrical and ever changing with moving reference point of perceiver as well as movement of sun over time. Object may be static but Nature-in the form of Sun with its intensity and angle of light varies with time over diurnal as well as annual cycles. This ever changing time-space alchemy of architectural space renders it experientially different all the time and thereby timeless. This is where good architecture transcends the symmetry and renders it experientially rich by its diversity and unpredictability through ever changing space perception.

It is just like in nature, where in most living organism the generic principle of symmetry is observed, but its rigidity, predictability and confines are broken by subtle nuances that it brings about some sense of wonder, anxiety, and sense of discovery. Like a face of human being, basic symmetrical structure is maintained but sutle variations are offered within the bilateral symmetry. Not perceptible evidently on its own it can be clearly when you mirror exactly either of the halves.

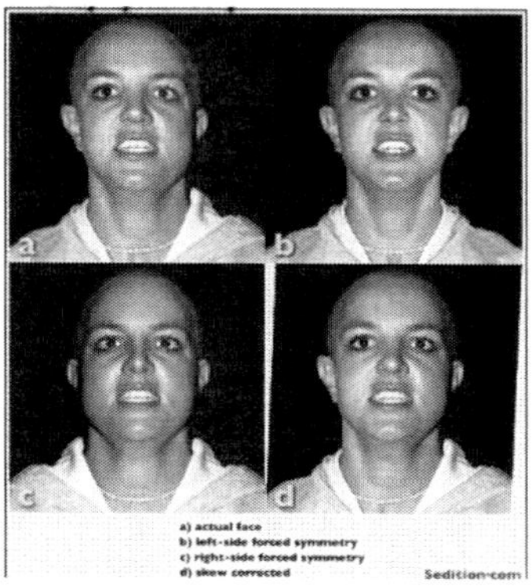

Movement is the key to spatial perception. Architecture is the story of movement and pauses where "kinaesthetics" of space is fundamental to its experience and perception. Architecture is a celebration of life. Manifestation of an idea, it encodes messages and emotes feelings. Architecture communicates through spatial tools, may they be the space sequences and its organization; elements of space making and their scale and form, or the symbolism of surface articulation. It is this aspect of encoding and decoding that sets in an instantaneous dialogue between the user and the architectural product. The affectivity of the communication depends on the easy comprehension of the encoded messages and their appropriate compliances in built form.

This communication takes place at three levels. Sensorial, experiential and associational. Sensorial perception refers primarily to physiological comforts arising out of physical conditions essentially in response to environmental control. For example small or large, dark or bright, cold or hot and so on. This bodily perception is humanly universal. Also universal is the experiential aspect; however it deals critically with the mind and cradles emotions. The process is spontaneous and reactionary to the nuances of space configuration and its dictates. While, the associational is the locale specific perception requiring pre-conditioning, familiarity or the acquired information base. For example symantics. It creates emotional bonds and succeeds through in depth understanding of cultural connotations. The complete communication is through wholesome balance of all the three.

'Kinesthetics' and temple architecture

A typical Hindu temple best illustrates this phenomenon. In plan temple assembly is organised symmetrically, but in reality it is perceived asymmetrically with changing sighlines and movement axis. In a temple, the sequence of gopurams, series of ascending steps and platforms, rising volumes of domes and shikharas, increasing degree of enclosure and the transition from the semi-open, multi directional pavilions to the unidirectional dark sanctum enclosed by solid walls, all heighten the progression from the corporeal to the spiritual as one progress from the gopuram to garbhagriha. This sense of transcendence from the worldly, from terrestrial to celestial, is further enhanced by the culmination of the horizontal planes of the platform into a vertical axis through tall pointed shikharas symbolically pointing towards the heavens. In this manner, the elements of a building, its scale, size, volume, degrees of enclosure, levels of illumination as well as motifs and decorations instil in the observer ethos appropriate to the place.

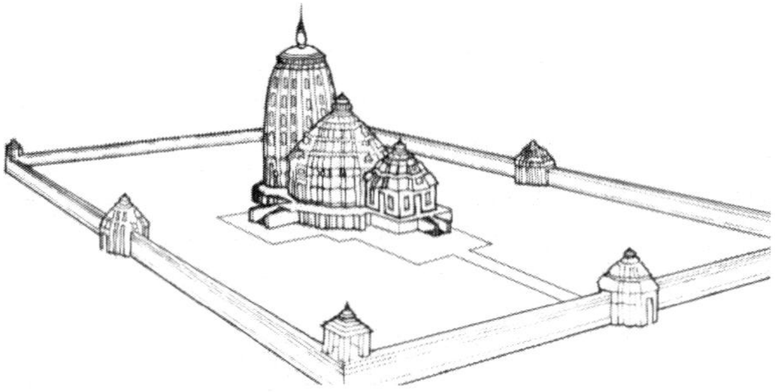

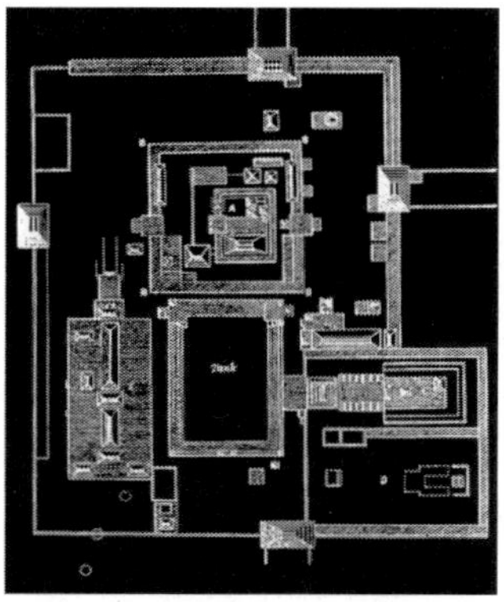

Hindu temple

Transcending time and space, good architecture remains communicative and interactive all the while through its spatial qualities. These spaces possess the qualities to establish rapport with the on looker and condition its perception, independent of their cultural background. Timeless, ever pervading architecture rely on more fundamental attributes of space making ranging from approach and movement, scale and proportion, quality of light and shade or the relationship of the built with the unbuilt. Such architecture therefore needs to be understood and interpreted through perceptual and experiential qualities and not by the abstractions of the plan geometry or static compositions of the facade elevations. Dynamics of moving through the space and sensory perception of it is vital to good architecture. [2]

It is for us to interpret spatial narratives built within the architectural space resolutions. There are innumerable nuances of built form by which traditional Indian Architecture manifests moods, communicates messages and remains relevant after these many centuries.

Dictums of space organisation for experientially engaging architecture

Some of these dictums of timeless and experientially driven spaces are:

(a) Non linear organization of spaces through shifting axis of movement which helps gradually unfold spaces and introduce element of surprise

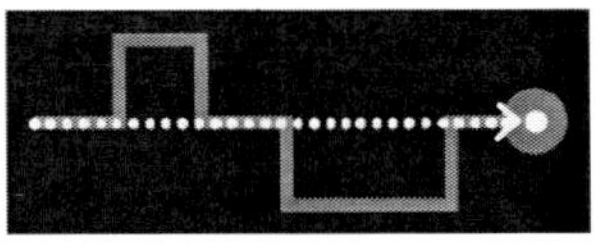

(b) The layering of spaces by baffles and thresholds induces an interactive process by building up the curiosity and the implicit sense of discovery

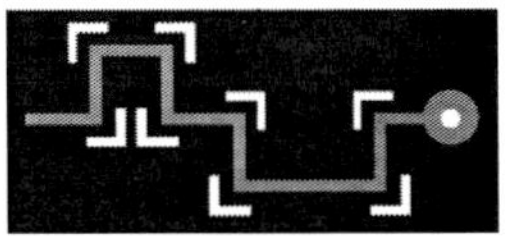

(c) The provision of pause points and thresholds help one reorient and reaffirm bearings in space. It provides clues and informed choices for individual preferences and personal pursuits.

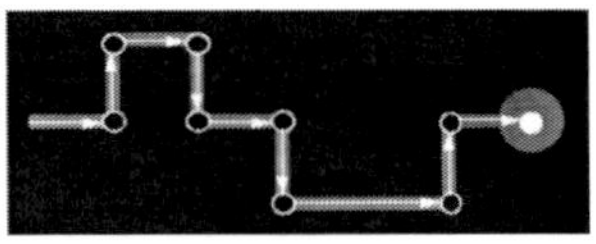

(d) element of time overlaid over space through increased path of movement and indirect approach, help condition mind and confront same space differently

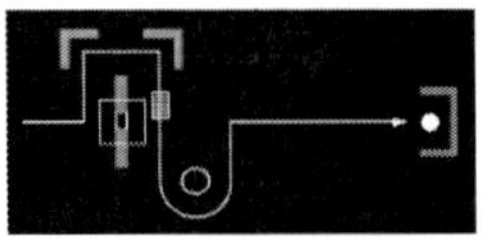

(e) The visual proportioning of spaces and skilful arrangement of elements provide ever-changing frames of visual compositions through perspectival effects and their strategic visual alignments while in movement

(f) Creative use of landscaping elements such as water and vegetation to enhance psychological and metaphoric overtones of the place and function. For example reflection of mosque dome in ablution pond revealed only after bending down for ablution help remind and condition mind for further goal.

(g) Play of light through integration of unbuilt space become mutual counterpoints help define space hierarchies and territories in terms of personal and public realms. The conditions of outdoors inherently not remaining same changes consistently making the same space appear different and fresh all the time.

These aspects of time over space lend varied meaning to the same space configuration through its conditioning, rendering architecture ever fresh, interactive and timeless.

The step wells of India illustrate the fact.

Adalaj stepwell and kinesthetic perception

Step well-water well with steps is a typology of building unique to western India. Essentially a device to contain water, it exalts to become a socio-religious institution – owing largely to its architectural manifestations! Spatially, the step well remains one of the most profound architectural statements. Despite its functional obsolescence of fetching water in today's context, it continues to inspire visitors even after six centuries. Transcending time and space it has rendered timeless, essentially through its spatial attributes and experience of moving through the same. These experiences are

Plan-Adalaj step well

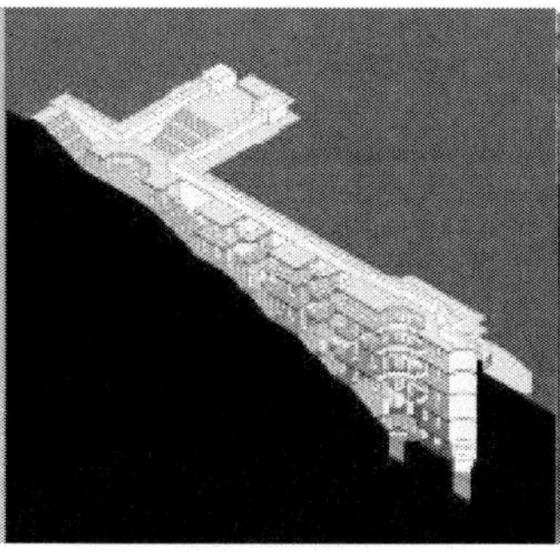

Cross sectional view-Adalaj step well

View-Adalaj step well

shared trans-culturally by all at a basic human plane without any familiarity with the context or any knowledge of its history. This also makes it universal where communication between the subject and the onlooker is inherently decipherable. It is the journey, the process of moving through the space which in itself becomes the event.

The clues for movement, inherent to the space are revealed sequentially. This gradual unfolding of spaces creates a sense of curiosity within the onlooker and involves them in the process. A dialogue is established between the subject and the onlooker through mutual process of encoding and decoding of messages and thus making the entire process interactive as well as the discovery very personal and intuitive.

At Adalaj step well, the only visible clue above the ground is the pair of pilasters flanking a wide flight of steps. They become the inviting portal to climb the steps leading to a platform. The plinth platform then leads the descent through the descending steps and retaining side walls.

First flight of steps descends to a square platform with an octagonal frame, what would once have been a dome base. This becomes the first point of pause. Surrounded on three sides by the inclined planes of rising steps the node leaves only one side open to approach. This is perpendicular to the axis of previous movement and thus involves a shift in visual as well as movement path. The new axis is marked by series of descending steps and intermittent platform located along the visual axis.

Although, a straight linear symmetrical organization of elements along the horizontal axis, the visual references continuously change due to inclined movement (through simultaneous displacement of horizontal and vertical axis) at every step. While the sight lines extend through its entire length the visual frame-composition of elements-constantly changes with the changing eye levels, vanishing points, perspectival alignments and the resultant visual composition all along its depth.

Each step reveals only the next while the subsequent flights of steps and water remain concealed by platforms from the cone of vision. Portals created by trabeated column and beam construction to shore the side walls retaining the earth are key spatial elements. The stone columns and beam provide the visual frames which enhance the sense of depth by providing an intermediate visual reference. With such changing visual frames the descent of five floors brings enface the actual water source.

As termination of the journey, steps flank the water on all four sides making the well the notional centre and the focus which connects vertically up to sky and down to water well.

This journey is enriched by the spatial variations emerging through sense of enclosure and intensity of light. The total descent at Adalaj is of five floors. At the first level it begins with ground and sky as the basic references while as one descends further, the walls begin to surface changing the perception of the degree of enclosure. While the first few levels still have reference to sky through skyward aperture, while the subsequent lower levels become more enclosed with only walls remaining in the cone of vision. The perception of enclosure is further reinforced by the decreasing intensity of light. The brightness of light subdues as one progresses further deeper into space making the space progressively vertical, cosier and confined. The timelessness is attributed to its freshness offered by the variations of visual frames as well as the integration of light/nature where the sun is the constant variable as it changes position and intensity from morning to evening and from summer to winter. This makes the static object change with changing outdoor conditions through sun.

The variation of visual frames illuminates the varying strengths of sun's light over time and season. It is to this dynamics we attribute the step well's universality. The journey of descent conditions the mind to an increasing sense of penetration in ground. The symbols along the path function as constant reminders of one's presence in the water world and creates the mood, ambiance and the mental frame such that finally reaching the water is as good as an ablution – a holy dip into the water. Thus through these notions and conditioning of mind, the well as a utilitarian object gets elevated to a shrine. Nourishing physically as well as spiritually it communicates beyond its functional and structural dictates. The sum total of the experience is sensorial as well as spiritual. It touches senses as well as mind. The meaning is in the journey itself and not necessarily the destination. The destination is not a means in itself, it is rather an excuse.

Thus, while symmetry may be a common place phenomena in the conception of space at the time of encoding the spatial syntax, especially in diagrammatic resolution of space in the two dimensional plane of PLAN layout; good architecture, driven experientially rather than by its form, transcends the symmetry by introducing the sense of layering, meander and sequential unfolding of spaces through kinetic perception of the viewer along the shifting axis and asymmetrical path of movement. This imbues sense of involvement, discovery and curiosity much needed to transform architecture in to an experience and process of decoding an event.

Major reference

[1] en.wikipedia.org/wiki/Symmetry (7th July, 2011)

[2] Concepts of space in Traditional Indian Architecture, Yatin Pandya, Mapin Publication (2005).

Symmetry: A Multi-Disciplinary Perspective No. 16, 2011, pp. 63–70.

Symmetry of Solutions of Differential Equations

Mythily Ramaswamy

T.I.F.R CAM, Sharada Nagar, Yelahanka New Town, Bangalore 560 065, India
e-mail: mythily@math.tifrbng.res.in

Abstract. The history of use of symmetry in differential equations via Lie group theory is sketched. Some instances of use of symmetry in finding special solutions of partial differential equations and in bifurcation are illustrated.

Keywords. Differential equations, group theory, Lie groups, radial symmetry, bifurcation.

1. Introduction

Symmetry considerations play an important role in the study of differential equations. In many natural phenomena, like atomic structure, crystal growth for example, symmetries are respected. Thus when a differential equation is derived as a mathematical model for a physical phenomenon, the underlying symmetries have to be taken into account. In fact, often the symmetries guide us in selecting the correct equation to model the phenomenon. These symmetries further aid in the resolution of the differential equation, analysis of the solutions and their numerical computations.

One of the important mathematical tools to study symmetry in various systems is group theory. The permutation groups were successfully used by Galois, to determine whether a polynomial equation was solvable by radicals or not. A more advanced group theoretic tool, namely the continuous transformation groups, introduced and developed by Sophus Lie, allowed one to carry out such a program for a differential equation, using the symmetry groups of that differential equation. Here we will see a brief outline of the historic developments, some simple instances of the use of symmetry in solving some partial differential equations, bifurcation and pattern formation.

2. Historical development

Some important questions in the theory of differential equations are the existence and uniqueness of solutions and their qualitative properties. Finding an exact solution of a differential equation is very useful but not always easy. This is hardly surprising since the polynomial equations are also not easy to solve. While the formula for the root of a quadratic equation was known for many centuries, solution of cubic and quartic equations were discovered much later around 16th century. Then it was a puzzle why a quintic equation was not solvable in this way. The reason why it is impossible was beautifully explained by Galois, around the year 1832. He showed that every polynomial equation could be associated with a group of permutations that represents the symmetry properties of that equation. This group is now known as Galois group. The equation is solvable only if the Galois group is "solvable". See for an interesting account of these developments, [5] the book by Mario Livio.

Lie was inspired by the work of Galois on the solutions of polynomial equations using permutation groups and he wanted to carry out a similar program for differential equations. This motivated him to develop the theory of continuous groups in 1872. Roughly speaking, these continuous transformation groups, now known as Lie groups, are groups with a "manifold" structure. The combination of algebraic techniques of group theory and differential geometric techniques contribute to the richness of Lie group theory.

The early developments in differential equations were motivated by a search for exact solutions. A variety of special techniques to solve certain types of equations – separable, homogeneous, exact, etc. were developed. This is what one studies in a first course in differential equations. It was Lie who observed that these are special instances of a general integration procedure using the invariance of the differential equation under a continuous group of symmetry. This unified and extended the available techniques for solving differential equations.

A symmetry group of a system of diffrential equations is a group that transforms the solutions of the system to other solutions of the same system. Some typical examples of such groups are translations, rotations and scaling. These symmetries arise from the structure of the differential equations and the symmetries of the domain. In order to find a symmetry group of the differential equation, one has to solve the equations arising from the conditions for invariance under the symmetry transformation. But these may also be complicated nonlinear equations. It is a profound discovery of Lie that the nonlinear conditions of invariance of the system under the continuous group of transformations could be replaced by equivalent simpler linear conditions, coming from "infinitesimal" invariance of the system under the "generators" of the group. This has the advantage that they can be found out by explicit computations. Once the symmetry group is known, that can be used to generate new solutions from known ones or to classify the solutions. See for example, [4], the book by Peter Hydon, Symmetry methods for differential equations, A beginner's guide or the more advanced book [6] by Peter Olver, Applications of Lie groups to differential equations.

In the case of ordinary differential equations (o.d.e's), Lie used the associated one parameter symmetry group to reduce the order of the differential equation. For a first order differential equation with a known one-parameter symmetry group, this leads to a solution via a single integration. In the case of multi-parameter symmetry groups, the order of the o.d.e may be reduced but the reduced equation may not be integrable unless further conditions hold. For partial differential equations (p.d.e's), the symmetry groups do not seem to help much in finding the general solution but the symmetry groups do help to determine special solutions which are invariant under some subgroup of the symmetry group of the system. These solutions can be found from a reduced equation, obtained by imposing the symmetry constraints, as in the case of o.d.e's. This is usually easier to solve. We will see some examples of this technique in the next section.

In some cases, the symmetry considerations are more useful, as in the case of the o.d.e arising from a variational principle. In such cases, Lie's techniques were further developed and used by E.Noether in her far reaching theorems, proved in 1918 relating the symmetry group of a variational integral with a conservation law for its associated Euler Lagrange equations.

In the years after 1893, E. Cartan gave an elegant abstract reformulation of differential geometry and Lie group theory. Then rapid developments in Lie group theory followed leaving behind the remaining difficult questions in differential equations.

G. Birkhoff in 1936, revived the applications of Lie group theory to differential equations of fluid mechanics. Important nonlinear p.d.e's like Korteweg-de Vries equation, offered pleasant surprises, such as the existence of infinitely many generalized symmetries and conservation laws in the spirit of the work of Noether. The application of symmetry group methods for finite dimensional Hamiltonian systems of o.d.e's have now been extended to Hamiltonian systems of evolution equations. See [6] for more details.

Symmetry considerations have also been very useful in understanding and predicting pattern formations. The term bifurcation roughly means the appearance of a new branch of solutions for a system of differential equations near a given branch of solutions. If the new solutions possess less symmetry than the original ones, then the system is said to have undergone symmetry breaking. Stability of the solution is another related important issue. Only the stable solutions are physically observable. Symmetry groups have been used to describe and classify the static as well as the dynamic patterns bifurcating from a given branch of solutions of a system of differential equations and to determine their stability. See for example [3].

Another obvious advantage in searching for solutions with some symmetry is that the numerical computations involved are considerably reduced. This has been implemented nowadays in many numerical algorithms to search for solutions of dynamical systems with symmetry.

3. Use of symmetry – Special solutions

Here we consider some simple examples of p.d.e's for which symmetry groups are used to write down some special solutions. One important use of symmetry considerations that we highlight here with examples, is the reduction in complexity of the differential equation and the resolution of the p.d.e. For more details, see the book by Evans, [1].

3.1 *Traveling wave solution*

Let us look for solutions with translation symmetry. Such solutions arise when there are traveling wave solutions – waves which translate themselves without change in shape. For a p.d.e of 2 real variables x and t, let us look for a traveling wave solution, with profile given by a function v of one variable, say, $v(x - \sigma t)$, traveling with speed $\sigma > 0$.

First we search for such solutions for the linear wave equation:

$$u_{tt}(x,t) - c^2 u_{xx}(x,t) = 0.$$

Searching for a solution of the form

$$u(x,t) = v(x - \sigma t),$$

we arrive at the relation $\sigma^2 - c^2 = 0$ or $\sigma = \pm c$. Thus the solution is

$$u(x,t) = u_1(x,t) + u_2(x,t)$$

where $u_1(x,t) = v(x - ct),\ u_2(x,t) = v(x + ct)$. This indeed represents a forward traveling wave u_1 as $u_1(x,0) = u_1(x + ct, t)$ and a wave u_2 moving backward.

Let us consider now the nonlinear Korteweg-de Vries equation, modeling surface waves in water,

$$u_t(x,t) + 6uu_x(x,t) + u_{xxx}(x,t) = 0$$

for $(x,t) \in \mathbb{R} \times (0,\infty)$. A search for traveling wave solution $v(x - \sigma t)$ leads to an o.d.e which gives after one integration

$$-\sigma v(s) + 3v^2(s) + v''(s) = \text{constant} = c.$$

After multiplying this equation by $v'(s)$ and integrating, we deduce

$$\frac{(v')^2}{2}(s) = -v^3(s) + \frac{\sigma}{2}v^2(s) + cv(s) + d$$

for some arbitrary constant d. We look for solutions satisfying

$$v(s), v'(s), v''(s) \to 0 \quad \text{as} \quad s \to \pm\infty.$$

Then the above equation reduces to

$$\frac{(v')^2}{2}(s) = v^2(s)\left(-v(s) + \frac{\sigma}{2}\right).$$

Hence we get $v'(s) = \pm v(s)(\sigma - 2v(s))^{1/2}$. Some more computations lead to an explicit solution

$$u(x,t) = \frac{\sigma}{2} sech^2\left(\frac{\sqrt{\sigma}}{2}(x - \sigma t - c)\right)$$

for each $c \in \mathbb{R}$, $\sigma > 0$. These special solutions are called Soliton solutions.

3.2 *Similarity solutions*

Another example of solution with symmetry is that of a solution invariant under dilation or scaling so that

$$u(x,t) = \lambda^{\alpha} u(\lambda^{\beta} x, \lambda t)$$

for all $\lambda > 0$, for some constants α, β. These are called self similar solutions. If we let $\lambda = \frac{1}{t}$, then the right hand side is $\lambda^{\alpha} u\big(\frac{x}{t^{\beta}}, 1\big)$. Thus the variable $y = \frac{x}{t^{\beta}}$ is important and is known as the similarity variable. If we call

$$v(y) = u(y, 1),$$

then we need to look for a solution of the form $\frac{1}{t^{\alpha}} v\big(\frac{x}{t^{\beta}}\big)$.

We consider first the linear heat equation,

$$u_t(x,t) - \Delta u(x,t) = 0$$

for $(x,t) \in \mathbb{R}^n \times (0,\infty)$. Here $\Delta u = \sum_{i=1}^{n} \frac{\partial^2 u}{\partial x_i^2}$. Looking for a solution $v(y)$ as above, we find for $y = \frac{x}{t^{\beta}}$

$$\alpha t^{-(\alpha+1)} v(y) + \beta t^{-(\alpha+1)} y \cdot \nabla v(y) + t^{-(\alpha+2\beta)} \Delta v(y) = 0.$$

Now a special choice of $\beta = 1/2$ allows us to reduce the above expression into one involving the variable y alone:

$$\alpha\, v(y) + \frac{1}{2}\, y \cdot \nabla v(y) + \Delta v(y) = 0.$$

A further simplification results, if we assume that v is radially symmetric:

$$v(y) = w(|y|)$$

so that the p.d.e reduces to an o.d.e in the radial variable. Now the o.d.e for $w(r)$ can be solved for a particular choice of $\alpha = \frac{n}{2}$. This leads us to a special solution for the heat equation:

$$u(x) = \begin{cases} \frac{1}{(4\pi t)^{n/2}} e^{-\frac{|x|^2}{4t}} & x \in \mathbb{R}^n, t > 0 \\ 0 & t < 0 \end{cases}$$

This is the most important solution, known as the fundamental solution of the heat equation. This is the building block to get solutions of other problems like inhomogeneous ones.

For the nonlinear porous medium equation,

$$u_t(x,t) - \Delta(u^{\gamma})(x,t) = 0$$

for $(x,t) \in \mathbb{R}^3 \times (0,\infty)$ for $u \geq 0$ and $\gamma > 1$, a constant, a search for such solutions leads to a relation

$$\alpha + 1 = \alpha\gamma + 2\beta$$

and a p.d.e in $y \in \mathbb{R}^3$. Thus we have reduced from 4 variables to 3 variables.

One can simplify further by searching for solutions with radial symmetry, that is

$$v(y) = w(|y|).$$

After solving the o.d.e for w in the radial variable, we get

$$u(x,t) = \frac{1}{t^{\alpha}}\left(b - \frac{\gamma - 1}{2\gamma}\beta\frac{|x|^2}{t^{2\beta}}\right)^{\left(\frac{1}{\gamma-1}\right)}$$

for some constant b. Here

$$\alpha = \frac{n}{n(\gamma - 1) + 2}, \qquad \beta = \frac{1}{n(\gamma - 1) + 2}.$$

This is known as Barenblatt's solution.

3.3 *Radially symmetric solutions*

Solutions which are invariant under rotations are functions of just one variable, the radial variable $r = |x|$, for $x \in \mathbb{R}^n$. Looking for such radially symmetric solutions is very useful as it reduces the p.d.e in n variables to an ode in the radial variable. This is what helped in the calculations in the previous section too. If the domain has rotational symmetry and if the differential equation is invariant under the action of the rotation group, then one can expect radial solutions. For example, when we look for a solution of the Laplace equation in the whole space, we look for

$$u(x) = v(|x|) = v(r)$$

so that $\Delta u = 0$ is transformed to

$$v''(r) + \frac{(n-1)}{r}v'(r) = 0$$

which is easily solved to get

$$u(x) = \begin{cases} b \log |x| + c & (n = 2) \\ \frac{b}{|x|^{n-2}} + c & (n \geq 3) \end{cases}$$

with constants b, c. For a special choice of the constants b and c, this gives the fundamental solution of the Laplace equation.

In some cases, we are able to say that all solutions, or at least all positive solutions, have radial symmetry. For example, in a ball B, the first eigenfunction $\phi(x)$ of the Laplacian, $(-\Delta)$ with zero Dirichlet boundary conditions on the boundary ∂B of B, satisfies

$$-\Delta\phi(x) = \lambda_1\,\phi(x) \quad \text{in } B,$$

$$\phi(x) = 0 \quad \text{on } \partial B.$$

This first eigenfunction $\phi(x)$ is known to be positive and radially symmetric. Here λ_1 is the first eigenvalue for the Laplacian with Dirichlet boundary conditions. In fact all positive solutions are radially symmetric even when we replace the linear right hand side λu by a smooth nonlinear function $f(u(x))$. This is an elegant result due to Gidas, Ni and Nirenberg stating that for a differentiable function f from $\mathbb{R}$ to $\mathbb{R}$, any positive solution of

$$-\Delta u(x) = f(u(x)) \quad x \in B,$$

$$u(x) = 0 \quad x \in \partial B,$$

is radially symmetric. (see [2]). This work has opened up an intense research activity exploring radial symmetry or lack of it for more general elliptic p.d.e's and less regular nonlinearities in different domains with symmetries. This also helps us to classify the solutions.

If we are able to prove further that the reduced o.d.e has a unique solution, then we can establish the uniqueness of positive solution to the above problem.

4. Use of symmetry in bifurcation

When new solution branches appear from existing solution branches, we call it "bifurcation". If the existing solution branch has some symmetry, then it is interesting to know what would be the symmetry of the new solution branch and which of these solutions are stable under perturbations. Bifurcation theory tells us when bifurcation would occur and what would be the structure of the new branch and also its stability properties. The main tool there is the linearized equation around a given solution. If that admits a nontrivial solution, then one can expect to prove bifurcation of a new branch of solutions. Symmetry group considerations tell us which of the subgroups of the original symmetry group can be the symmetry group for the bifurcating branch of new solutions. See for example, equivariant branching lemma in [3] for bifurcation of solutions invariant under a symmetry group.

An interesting example is Taylor–Couette flow in a fluid contained in between 2 concentric cylinders outer one stationary and inner one rotating with constant speed c. As we increase this speed, the initial flow, known as Couette flow, gives rise to Taylor vortices, spiral vortices and wavy vortices and then turbulence sets in. (See some pictures of the patterns observed in experiments in [3], section 6.1).

The governing equations are the famous Navier-Stokes equations for the velocity vector of the fluid in a cylindrical domain with periodic boundary conditions. The Navier-Stokes equations are invariant under the Euclidean group E_3 consisting of translations, rotations and reflections of space. The domain has rotational symmetry about the axis of the cylinder and translational symmetry along vertical axis and reflectional symmetry about a plane perpendicular to the axis of the cylinder. The full symmetry group of this model is $O(2) \times SO(2)$. Here $O(2)$ is the orthogonal group on $\mathbb{R}^2$ given by 2×2 matrices A such that $AA^T = I$ and $SO(2)$ is the special orthogonal group given by matrices A with the additional condition that the determinant is 1.

Couette flow is the steady state observed in experiments for low speeds. The first bifurcation from this steady state is to Taylor vortices, which looks like stacked "vada's". Couette flow is invariant under the full symmetry group but the new solution bifurcating from it, has lost some of the symmetry. This is indeed a symmetry breaking bifurcation. The new solution still is invariant under some subgroup of the original symmetry group. In the case of Taylor vortices, it has lost the translational symmetry along the vertical axis but it still has the rotational symmetry. As the speed is increased, the Taylor vortices undergo further bifurcations. In each of these bifurcations, some of the symmetry of the existing solution is broken and a certain smaller subgroup of symmetry is selected for the new bifurcating solution. See for details, the analysis in [3].

In recent years, symmetry in nonlinear dynamical systems has been actively studied to understand, analyze and predict static as well as dynamic behaviour in various scientific disciplines. Some examples are time periodic solutions via Hopf bifurcation, heteroclinic cycles, symmetric chaos, periodic solutions of symmetric Hamiltonian systems etc. A readable account of these developments is given in [3].

References

[1] Evans, Lawrence C. Partial differential equations, Graduate Studies in Mathematics, 19. American Mathematical Society, Providence, RI (2010).

[2] Gidas, B.; Ni, Wei Ming; Nirenberg, L. Symmetry and related properties via the maximum principle, *Comm. Math. Phys.*, **68** (1979) no. 3, 209–243.

[3] Golubitsky, Martin; Stewart, Ian The symmetry perspective. From equilibrium to chaos in phase space and physical space. Progress in Mathematics, 200. Birkhäuser Verlag, Basel (2002).

[4] Hydon, Peter E. Symmetry methods for differential equations. A beginner's guide. Cambridge Texts in Applied Mathe matics. Cambridge University Press, Cambridge (2000).

[5] Livio, Mario, The equation that couldn't be solved. How mathematical genius discovered the language of symmetry. Simon & Schuster, New York (2005).

[6] Olver, Peter J. Applications of Lie groups to differential equations. Second edition. Graduate Texts in Mathematics, 107. Springer-Verlag, New York (1993).

Symmetry: A Multi-Disciplinary Perspective No. 16, 2011, pp. 71–95.

Symmetry in Molecular Structure and Dynamics

Srihari Keshavamurthy

Department of Chemistry, Indian Institute of Technology, Kanpur 208 016, India
e-mail: srihari@iitk.ac.in

Abstract. In this chapter I illustrate the power of symmetry in certain areas of chemistry. Applications are given both in structure and dynamics. In particular, the examples of fullerene, crystal field theory and the Huckel model for conjugated molecules clearly brings out the utility of symmetry principles in structure. The role of spins, indistinguishability, and permutations in chemical dynamics are illustrated with appropriate examples. A glimpse, at a rather basic level, into topics such as the Coulson-Rushbrooke pairing theorem and the complete nuclear permutation inversion groups is also provided.

Keywords. Symmetry, fullerene, icosahedral, crystal field, graphs, bichromatic, permutations, CNPI.

1. Introduction

Symmetry is a very powerful concept. The origins of symmetry principles go back several centuries. Perhaps it is appropriate to start with a quote from the beautiful book [1] on symmetry by the great Hermann Weyl: "*Symmetry, as wide or as narrow as you may define its meaning, is one idea by which man through the ages has tried to comprehend and create order, beauty, and perfection*". Indeed, there are many levels at which we recoginize, appreciate, and exploit symmetries in nature. The easy examples are the ones that are visible to us and nature provides the best examples. Thus, the beauty of a crystal, the delicate patterns of the snow flakes, the arrangements of the petals on various flowers and even the veins on a leaf are all obviously pleasing to our sensory perceptions. The sense of awe when we see the architectural beauty of the Hoysala temple in Belur, the Taj Mahal in Agra or the Sagrada Familia in Barcelona is again due to our innate appreciation for all things symmetrical and beautiful. Indeed even a well presented platter of food delicacies seems to make us more hungry when compared to the same food being presented rather sloppily! I might add here that art (in all its variations) and music (in every form) come with their own symmetries and with some effort all of us appreciate the nuances in such creative fields. The symmetries in art and music, however, require some effort on our part since they embody the symmetries at a slightly abstract level. Incidentally, the great composer Igor Stravinsky once said that "*to be perfectly symmetrical is to be perfectly dead*". What does that mean? Does one have such notions in the physical sciences as well? That should become apparent from some of the other lectures in this book. The relevance to chemistry will become clear later on in this chapter.

This brings us to the subject of symmetries which are not obviously perceived by our senses or those that are "hidden" from us. For instance, here is an identity written down by the legendary Indian mathematician Ramanujan [2]

$$\sqrt{\alpha}\int_0^\infty \frac{e^{-x^2}}{e^{\alpha x}+e^{-\alpha x}}dx = \sqrt{\beta}\int_0^\infty \frac{e^{-x^2}}{e^{\beta x}+e^{-\beta x}}dx \tag{1}$$

which is true not only for the trivial case of $\alpha = \beta$ but also for the nontrivial case of $\alpha\beta = \pi$. Is this symmetry? Specifically, is this symmetry comparable to that of the snow flakes or the architectural wonders? Apparently, the answer is a resounding yes. Thus, G. N. Watson who studied several of the Ramanujan identities in great detail remarked that many of the identities evoke emotions that are identical to that when he enters the Sagrestia Nuova of Capelle Medicee and sees the sculptures of Michelangelo set over the Medici tombs. Closer home, here are the famous Maxwell's equations of electrodynamics in vacuum:

$$\nabla \cdot \boldsymbol{E} = 0 \tag{2}$$

$$\nabla \cdot \boldsymbol{B} = 0 \tag{3}$$

$$\nabla \times \boldsymbol{E} = -\frac{1}{c}\frac{\partial \boldsymbol{B}}{\partial t} \tag{4}$$

$$\nabla \times \boldsymbol{B} = +\frac{1}{c}\frac{\partial \boldsymbol{E}}{\partial t} \tag{5}$$

Even without knowing that $\boldsymbol{E}$ and $\boldsymbol{B}$ are the electric and magnetic fields, you would be able to appreciate the symmetry between the four equations above. Note the subtle sign change in the last pair of equations. The symmetry however is not perfect since magnetic monopoles do not exist (atleast we have not been able to detect them until now). So, here we see a hint about nature being not perfectly symmetrical (remember Stravinsky's remark made earlier).

On a slightly different note, here is an example where the symmetry is not in the equation but in what the equation generates. Suppose one plays a game where you start with a complex number $z_1 = \mathrm{C}$ and then square it and add C to generate $z_2 = \mathrm{C}^2 + \mathrm{C}$ and continue the process to generate the sequence $\{z_1, z_2, z_3, \ldots\}$ i.e., consider the iterative equation

$$z_{n+1} = z_n^2 + \mathrm{C} \tag{6}$$

Every choice of C corresponds to a point on the complex plane. Now if we plot only those points, and a large number of them, that do not escape off to infinity then we will get an intricate figure – the so called Mandelbrot set [3]. Amazingly, the resulting figure has a very special kind of symmetry and also very pleasing to look at. Does this have anything to do with physical systems or is it just a mathematical game? Turns out that the simple iterative game above is related to the notion of fractals which are scale invariant. So, you get the same figure as you keep on zooming on some portion of the original figure. The scale invariance property shows up in many physical systems [4].

Not surprisingly, symmetry plays a crucial role in chemical structure and dynamics which is the focus of this chapter. Here again there are several opinions and viewpoints on the utility of symmetry. I like the comment [5] by Roald Hoffmann in his review of the book [6] by Heilbronner and Dunitz – "*Symmetry is beautiful (in a simple way), symmetry elicits a sense of peace, repose, and stability. But it's really asymmetry that's interesting. In asymmetry is variety and tension. And richness. Perhaps I'm not entirely fair. Asymmetry without an inkling of order is chaos. Whatever beauty is, the tense edge of symmetry and asymmetry contributes to it.*" What exactly does Hoffmann mean by the tense edge of symmetry and asymmetry? Actually this tense edge exists in every physical science, perhaps even in the mathematical sciences. For example, there are several examples in biology, including our own body's approximate bilateral symmetry, wherein nature thrives on this tense edge between symmetry and asymmetry. In chemistry the whole field of asymmetric synthesis is all about embedding a specific chirality into the final product by using an asymmetric catalyst that breaks the symmety between pathways leading to the enantiomers. And this product might be a life saver; here respecting the symmetry of the pathways would be fatal, literally. In this context it is hard not to mention the ultimate symmetry breaking – the arrow of time. However, in chemistry there are also several examples where nature likes to restore symmetry instead of losing or lowering it. The prime example being the phenomenon of tunneling [7]. Tunneling can be viewed, in some sense, as the supreme symmetrizer and the process itself leads to very interesting dynamics in molecular systems.

There are several books [8] and chapters that have been written on chemical applications of symmetry and group theory. Consequently, it is a nontrivial task to write another chapter on the same theme with a new or fresh outlook. Still, in this chapter I have tried to present things from a slightly different perspective with emphasis on the general principles of symmetry. I have also attempted to present some topics, certainly advanced from the undergraduate point of view, which are usually not included in elementary texts. The purpose ofl giving a brief introduction to such topics is to convey the message that symmetry principles are not only important from a structurual point of view but also crucial in order to understand and manipulate chemical dynamics. Hopefully, the next few sections will amply illustrate the power of symmetry in molecular structure and dynamics.

2. C_{60}: The beautiful molecule

Every student of chemistry knows about the fullerene story. Richard Smalley, one of the three awarded the nobel prize for the discovery of C_{60} says [9] that "*the discovery was one of the most spiritual experiences that any of us in the original team of five have ever experienced*". Harry Kroto, the other winner, says [10] in his nobel lecture that "*There had been no premeditated research aim, the original reason for constructing the giant fullerene model was solely for the intrinsic interest and pleasure of building an elegant structure*". Incidentally, Kroto's lecture title is 'Symmetry, Space, Stars and C_{60}' and that just might give you an idea of how revered this molecule is!

Actually, fullerene has symmetry at several levels that one can appreciate. The first is the obvious visually pleasing effect – it is hard not to be enraptured by it. Secondly, as we shall see below, the character table of C_{60} has numbers related to the famous golden ratio, $\gamma \equiv (\sqrt{5}+1)/2$, found in various other natural structures [11]. The structure of C_{60} is that of a truncated icosahedron and is an example of an Archimedean solid. It has 32 faces with 20 being hexagonal and 12 being pentagonal. Actually, you need the pentagons since you cannot simply curl up a graphite sheet into the football shape of C_{60}. It is interesting that C_{60} is almost round but not perfectly round. One may ask the question as to what would happen if it was perfectly round. In this case group theory has several interesting lessons for us to learn. So, as a very simple model, imagine C_{60} to be a sphere and further assume that the 60 π-electrons are noninteracting. The quantum mechanical time-independent Schrödinger equation for a particle on a sphere is an easily solvable system. For a single particle, electron in this case, the equation is

$$-\frac{\hbar^2}{2m_e}\left(\frac{\partial^2}{\partial x^2}+\frac{\partial^2}{\partial y^2}+\frac{\partial^2}{\partial z^2}\right)\Psi(x,y,z)=\mathrm{E}\Psi(x,y,z) \tag{7}$$

Realizing that the spherical polar coordinates (r,θ,ϕ) are the most natural for the problem at hand one can use

$$\begin{aligned} x &= r\sin\theta\cos\phi \\ y &= r\sin\theta\sin\phi \\ z &= r\cos\theta \end{aligned} \tag{8}$$

to transform the partial differential equation into

$$-\frac{\hbar^2}{2m_e r^2}\left[\frac{1}{\sin\theta}\frac{\partial}{\partial\theta}\left(\sin\theta\frac{\partial}{\partial\theta}\right)+\frac{1}{\sin^2\theta}\frac{\partial^2}{\partial\phi^2}\right]\Psi(\theta,\phi)=\mathrm{E}\Psi(\theta,\phi) \tag{9}$$

Note that the radius r is a constant and hence the wavefunction only depends on the angles. The above differential equation is well known and yields the spherical harmonics i.e., $\Psi(\theta,\phi)\equiv \mathrm{Y}_{\mathrm{LM}}(\theta,\phi)$ with $\mathrm{L}(=0,1,2,\ldots)$ and $\mathrm{M}(=-\mathrm{L},-\mathrm{L}+1,\ldots,\mathrm{L}-1,\mathrm{L})$ being the two quantum numbers. These quantum numbers must be familiar to you from the Hydrogen atom example. The quantized energies that each electron can take are determined from

$$\mathrm{E}_{\mathrm{LM}}=\frac{\hbar^2}{2m_e r^2}\mathrm{L}(\mathrm{L}+1) \tag{10}$$

One has to be careful and take into account the $(2\mathrm{L}+1)$ fold degeneracy of the energy levels for a given L. Upon filling the energy levels using the Pauli exclusion principle, no more than two electrons per level, we find that the $\mathrm{L}=5$ level (called as the highest occupied molecular orbital or HOMO) is only partially filled. Infact, this 11-fold degenerate level has unpaired electrons. However, C_{60} is known to be an insulator and the species K_3C_{60} is a conductor. Is our assumption of a spherical buckyball consistent with these facts? No, it is not.

So, how does one explain and understand the properties of C_{60}? Turns out that it is important to realize that the bucky is almost round but not perfectly round. The actual symmetry of relevance here is the so called icosahedral group or I_h in the standard notation. Compared to the highly symmetric sphere, the truncated icosahedron is less symmetric. This feature is key since it leads to a very different energy level structure as compared to what we had for the sphere. So, to begin with, let us see some of the features of the icosahedron group.

2.1 *The icosahedral group*

The correct way to describe the symetry of a molecule is to make a list of all the symmetry operations that leave the molecule invariant. In molecular systems one is typically interested in the so-called discrete point groups. The symmetry elements are usually determined by studying the molecule under various rotations about different axes, reflections through planes containing such axes and inversion through the centre of symmetry [8]. My intention, and this is certainly outside the scope of these notes, is not to provide an introduction to mathematical group theory. Instead, I will assume some basic knowledge of groups and highlight or use the key features of certain groups. In the present case the elements that make up the icosahedral group can be expressed as

$$\mathrm{I}_h = (\mathrm{E}, 12\mathrm{C}_5, 12\mathrm{C}_5^2, 20\mathrm{C}_3, 15\mathrm{C}_2, i, 12\mathrm{S}_{10}, 12\mathrm{S}_{10}^2, 20\mathrm{S}_6, 15\sigma) \tag{11}$$

and hence a total of 120 symmetry elements. One says that I_h is a group of order 120 and several elegant theorems in group theory can be invoked to make predictions about the structure and spectroscopy of the molecule. For now let us focus on what happens to the $\mathrm{L} = 5$ degenerate energy level of the spherical C_{60}.

Degeneracy is associated with symmetry. So, whenever symmetry is modified then we should expect degeneracies to change. For example, the well known $2n^2$ degeneracy of the atomic energy levels of the hydrogen atom is due to the every high symmetry; in this instance it is due to the "hidden" symmetry which makes the energy levels depend only on one of the quantum numbers – the prinicipal quantum number (n). If one adds a static electric field in the z-direction then the symmetry is lowered and so is the degeneracy. Another elementary example is that of a single spin-1/2 particle like electron. In the absence of an external magnetic field the up ($+1/2$) and down ($-1/2$) projections are degenerate. However, the degeneracy is lifted with an external magnetic field and this simple fact is at the heart of the principle of NMR. Note that, as in the electron spin example, the degeneracies might even completely disappear. In the case of C_{60} we have lowered the symmetry from that of a sphere to the truncated icosahedron. So, the 11-fold degenerate HOMO in the previous case may "split-up" into many energy levels. The question is as to how many levels does the $\mathrm{L} = 5$ level split up into and in this case the properties of the group I_h allow us to determine this very precisely. Specifically, the character table of I_h is required at this stage.

The character table, as the name suggests, is the unique identity (character) of a group. The character table is of great utility in chemistry and therefore is given

Table 1. The character table for the icosahedral group I_h.

I_h	E	$12C_5$	$12C_5^2$	$20C_3$	$15C_2$	i	$12S_{10}$	$12S_{10}^2$	$20S_6$	15σ
A_g	1	1	1	1	1	1	1	1	1	1
T_{1g}	3	γ	$\bar{\gamma}$	0	-1	3	$\bar{\gamma}$	γ	0	-1
T_{2g}	3	$\bar{\gamma}$	γ	0	-1	3	γ	$\bar{\gamma}$	0	-1
G_g	4	-1	-1	1	0	4	-1	-1	1	0
H_g	5	0	0	-1	1	5	0	0	-1	1
A_u	1	1	1	1	1	-1	-1	-1	-1	-1
T_{1u}	3	γ	$\bar{\gamma}$	0	-1	-3	$-\bar{\gamma}$	$-\gamma$	0	1
T_{2u}	3	$\bar{\gamma}$	γ	0	-1	-3	$-\gamma$	$-\bar{\gamma}$	0	1
G_u	4	-1	-1	1	0	-4	1	1	-1	0
H_u	5	0	0	-1	1	-5	0	0	1	-1

in almost every book on physical chemistry. Essentially, such a table lists all the irreducible representations (irreps) of the group. With some effort one can construct the character table for any group. The character table for the icosahedral group is given below. One other aspect of the beauty of this group is hinted by the fact that the golden ratio γ shows up in the table[1]. I will not take you through the process here but show you how to determine the dimensionalities of the irreps as an example. If you look carefully at the way that the group elements have been shown above then you can see that all rotations C_n with specific n have been grouped together. Similarly, all the reflection planes σ have been grouped together. These are called as the classes and I_h has 10 classes. This in turn implies that there are 10 irreps, denoted as Γ_i, of I_h with possible dimensionalities d_i. A theorem in group theory tells us that there is a relation between the d_i and the order of the group

$$d_1^2 + d_2^2 + \cdots + d_{10}^2 = 120 \quad (12)$$

We can simplify the above by realizing that the dimensions must come in pairs due to the presence of the inversion symmetry. In other words if we find one 2-dimensional irrep then there must be automatically one more with the same dimensionality. Moreover every group has one so-called totally symmetric irrep of dimensionality one. Thus, we can write

$$d_1^2 + d_2^2 + d_3^2 + d_4^2 = \frac{120}{2} - 1 = 59 \quad (13)$$

and only four of the dimensionalities need to be determined. Remember, the d_i have to be integers and zero is not a physical choice in our case. So, can the number 59 be represented as the sum of four squares? Amazingly enough, Lagrange had already proved back in 1770 that no more than four squares are ever needed to express any number, irrespective of how large that number might be. Therefore, we are guaranteed of a solution to the above equation. Indeed,

$$3^2 + 3^2 + 4^2 + 5^2 = 59 \quad (14)$$

[1]Amongst the various point groups that are tabualted in group theory textbooks, γ shows up only in the I_h character table. Moreover, the series $\gamma + \bar{\gamma}$, $\gamma^2 + \bar{\gamma}^2$, $\gamma^3 + \bar{\gamma}^3, \ldots$ generates the Fibonacci series!

and hence we see that the group I_h has five pairs (one each tagged with the 'g' for gerade and 'u' for the ungerade) of irreps with dimensionalities 1, 3, 3, 4, and 5. The group theory notation for these irreps are A, T_1, T_2, G and H respectively.

2.2 *Consequence of I_h for C_{60}*

The consequence of the above discovered dimensionalities of the irreps of I_h is that the L = 5 eleven-fold degenerate energy level must split up into energy levels which can only have dimensionalities 1, 3, 3, 4, and 5. Thus, there is no possibility of having a doubly degenerate split level. From a purely number theoretic perspective we are interested in the number of ways in which the integer 11 can be expressed using the integers 1, 3, 4, and 5. Turns out that there are fourteen different ways[2]. So, there is more then one choice and we now do have to appeal to group theory to identify the correct choice. In particular, the correct choice among the possible fourteen comes about by looking at how the L = 5 spherical harmonics transform under the rotations of the icosahedral group. Why rotations? This is because on going from the sphere to the icosahedron one has now replaced the full rotation (continuous) symmetry by the finite rotation symmetry expressed via the various C_n. Thus we will be only interested in the group operations (E, C_5, C_5^2, C_3, C_2). The essential information comes from the character χ_n corresponding to a C_n rotation and given by the formula

$$\chi_n = \frac{\sin[(2L+1)\pi/n]}{\sin[\pi/n]} \tag{15}$$

Remember that the identity is a 'do nothing' operation and hence the corresponding character $\chi(\mathrm{E}) = 11$. Thus, we get the characters $(11, 1, 1, -1, -1)$ corresponding to the operations (E, C_5, C_5^2, C_3, C_2) respectively. We will denote this set of integers by Γ and looking at the character table for I_h one immediately notices that it is missing. This suggests that Γ that we have at hand is a 'reducible' representation. In other words, since the character table for I_h exhibits the irreducible representations, our reducible Γ can be expressed in terms of the irreducible representations[3] It is useful to note that $\chi(\mathrm{E})$ above is the character of the identity operation in the reducible representation. In the irreducible representation, for instance T_{1g}, we have $\chi_{T_{1g}}(\mathrm{E}) = 3$ which is different from $\chi_\Gamma(\mathrm{E}) = 11$. What we need at this stage is to determine the extent of contribution c_k of the various irreps χ_k of the icosahedral group. Mathematically, we can write

$$\chi_\Gamma(g) = \sum_k c_k \chi_k(g) \tag{16}$$

---

[2]The general problem of the total number of integer partitions $p(n)$ of a given integer n is an interesting number theory problem in itself [12]. Hardy and Ramanujan were the first ones to provide a very accurate estimate for this using some very neat mathematical tools. In any case, $p(11) = 56$ and we have far fewer in our case since partitions like $6+3+1+1$ are not allowed as the group I_h has no six dimensional irreps. Similarly any partition involving the ineteger 2 is not allowed for our case.

[3]It is instructive to think of the analogy between this procedure and the notion of expressing an arbitrary vector **v** in terms of the unit basis vectors $(\hat{i}, \hat{j}, \hat{k})$. So, in this analogy $\mathbf{v} = \hat{i}v_x + \hat{j}v_y + \hat{k}v_z$ and therefore **v** is reducible in terms of the irreducible unit vectors. Note that the dimensionality of the space here is three due to the fact that we need three basis vectors to express any arbitrary vector in this space.

for a specific element g of the I_h group. Using the orthogonality of the irreps the above equation yields

$$c_k = \frac{1}{O(I_h)} \sum_{g \in I_h} \chi_\Gamma(g)\chi_k(g) \tag{17}$$

with $O(I_h) = 120/2 = 60$ being half the order of the group. One is using half the order here since the relevant group operations are just (E, C_5, C_5^2, C_3, C_2). To see how this works let us calculate to see if the totally symmetric irrep A makes any contribution to our reducible Γ. The calculation proceeds as follows:

$$\begin{aligned} C_A &= \frac{1}{60}[11(1)(1) + 12(1)(1) + 12(1)(1) + 20(-1)(1) + 15(-1)(1)] \\ &= 0 \end{aligned} \tag{18}$$

A few more such calculations are

$$C_T = \frac{1}{60}[11(1)(3) + 12(1)(\gamma) + 12(1)(\bar{\gamma}) + 20(-1)(0) + 15(-1)(-1)] = 1$$

$$C_G = \frac{1}{60}[11(1)(4) + 12(1)(-1) + 12(1)(-1) + 20(-1)(1) + 15(-1)(0)] = 0$$

Thus, performing the calculations one finds that only H, T_1, and T_2 irreps are enough to represent Γ. In group theory, and mathematically more precisely, we should write

$$\begin{aligned} \Gamma &= H \oplus T_1 \oplus T_2 \\ &= H_u \oplus T_{1u} \oplus T_{2u} \end{aligned} \tag{19}$$

Notice that in the last line above I have somehow magically appended the 'u' tag on the irreps. There is of course no magic here – since we are examining the set of levels with $L = 5$, they are odd and hence the tags better be as shown. The symbol $\oplus$ is the so-called direct sum.

So, using the group character table we see that the only allowed, physically that is, partition of 11 is $5+3+3$ and neither the one or four dimensional irreps are involved. Thus, the original 11-fold degenerate $L = 5$ levels split up into a 5-fold degenerate and two 3-fold degenerate levels due to the breaking of the full rotational symmetry to the icosahedral symmetry. However, and this is a very crucial point, our analysis does not reveal the energetic ordering of the new levels. For this we have to work harder. Such a calculation will not concern us here, but a few experimental facts can lead us to the correct ordering. Firstly, the electrons must be all paired up for reasons mentioned earlier. Secondly, experimentally it is found that K_3C_{60} is a conductor while K_6C_{60} is an insulator. A little bit of thought shows that the simple model that we have will fit the facts only if the 5-fold degenerate H_u level is energetically the lowest. Even with all this information we cannot guess as to whether the T_{1u} or the T_{2u} will be the next in the energy ordering. At this stage we have to calculate appropriate quantum matrix elements and then one would find that

$$E_H < E_{T_1} < E_{T_2} \tag{20}$$

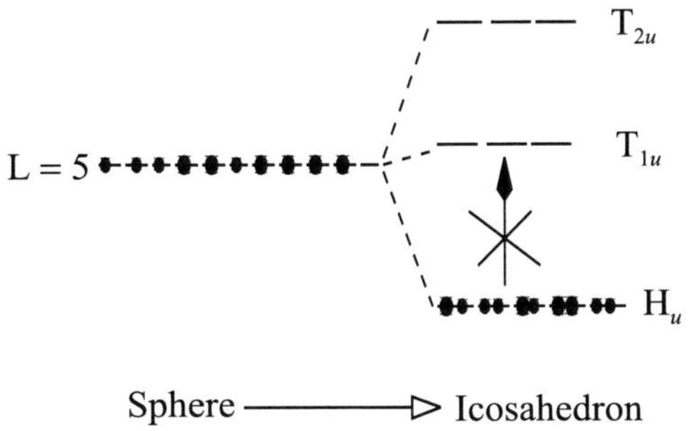

Figure 1. Splitting of the L = 5 degenerate states due to the icosahedral structure of fullerene. Group theory also tells us that the $H_{1u} \rightarrow T_{1u}$ optical transition is forbidden (illustrated by a crossed out arrow). Note that group theory does not tell us about the relative ordering of the split energy levels.

This example, therefore, shows the power of group theory and symmetry arguments to help understand and rationalize the experiments. One can discover many other facts using the elementary group theoretic techniques above. For instance, it is possible to show that the $H_{1u} \rightarrow T_{1u}$ optical transition is not allowed. Consequently, to measure this energy gap one would have to use different spectroscopic methods. Similarly, it is easy to discover the infrared and Raman active vibrational modes of C_{60} and, with some effort, a detailed understanding of the electronic spectrum. For further information along these lines I refer you to the original literature [13].

3. Crystal field theory: Utility of symmetry principles

As another example for the power of group theory and symmetry principles we will consider a metal atom with a single d-valence electron surrounded by six point charges of magnitude Q in an octahedral arrangement[4]. In other words, the central metal atom is under the influence of the electrostatic field produced by the point charges placed at $(\pm a, 0, 0)$, $(0, \pm a, 0)$ and $(0, 0, \pm a)$. We are using cartesian coordinate system. In this example, which is probably familiar to you from coordination chemistry, the 10-fold degeneracy of the d-electron is broken due to the symmetry being lowered from that of full rotation to an octahedral one. This leads to the famous $t_{2g} \oplus e_g$ splitting and can be worked out in analogy to the previous discussion on fullerene. However, I would like to illustrate a different aspect of the problem. This has to do with the form of the electrostatic field at the metal atom. First, I will take the "hard" approach wherein we will use explicit expressions for the electrostatic potential from the six point charges and sum them up to find the effective potential $V_c(\mathbf{r})$ at the metal atom. Then, we will derive the same result in a relatively easier fashion using group theory.

[4]We are neglecting the covalent effects and hence simplifying the ligand charge distributions to a point charge model. This results in the so called crystal field approximation. In order to explain the bonding and magnetic properties of metal-ligand systems one must appeal to the ligand field theory. Crystal field theory was first introduced by Hans Bethe in his pioneering 1929 paper [14]. Ligand field theory has it origins in the remarkable 1935 paper [15] of John Hasbrouck van Vleck.

The potential at the atom, placed at the origin of our coordinate system, due to the charge at $(a, 0, 0)$ is given by

$$\frac{Q}{\sqrt{(x-a)^2+y^2+z^2}} = \frac{Q}{\sqrt{r^2-2ax+a^2}} \tag{21}$$

with (x, y, z) being the location of the lone electron and we have denoted $r^2 = x^2 + y^2 + z^2$. However, since the charges and the metal atom are well separated, we have $r \ll a$ and hence we can expand the potential as

$$\begin{aligned}\frac{Q}{\sqrt{r^2-2ax+a^2}} &= \frac{Ze^2}{a}\left[1-\frac{2x}{a}+\frac{r^2}{a^2}\right]^{-1/2} \\ &= 1-\frac{1}{2}C+\frac{3}{8}C^2-\frac{5}{16}C^3+\frac{35}{128}C^4-\cdots\end{aligned} \tag{22}$$

with

$$C \equiv -\frac{2x}{a}+\frac{r^2}{a^2} \tag{23}$$

Notice that a similar expansion for the potential due to the charge at $(-a, 0, 0)$ can be obtained from the above by simply replacing $a \to -a$. Thus, keeping terms up to order four we have

$$\begin{aligned}&\frac{1}{\sqrt{(x-a)^2+y^2+z^2}} + \frac{1}{\sqrt{(x+a)^2+y^2+z^2}} \\ &\approx 2-\frac{1}{a^2}(3x^2-r^2)+\frac{1}{4a^4}(3r^4-30x^2r^2+35x^4)\end{aligned}$$

The contributions from the charges along the other two axes can be worked out similarly (or just by noticing the symmetry!) and summing it all up we get the effective potential felt by the electron as

$$V_c(\mathbf{r}) \approx Q\left[\frac{6}{a}+\frac{35}{4a^5}\left(x^4+y^4+z^4-\frac{3}{5}r^4\right)\right] \tag{24}$$

Remember that the above is approximate in the sense that we have retained only terms upto a certain order. One can certainly go to higher orders and work out the additional terms, but that requires some more effort. For what I need to discuss it is sufficient to keep things at this order. You may wonder as to why I did not keep only upto second order to begin with – surely, a look at the expression above will tell you the answer!

Ok, so how can the above derivation of the crystal field potential be obtained from group theoretic methods? To begin, note that $V_c(\mathbf{r})$ must respect the symmetry of an octahedron. The octahedral group O_h has order 48 and the symmetry operations of this group either flip the sign on one of the components of $\mathbf{r}$ (for example, $x \to -x$) or exchange (for example, $x \to y$) them. So, the statement that $V_c(\mathbf{r})$ must respect the symmetry of an octahedron implies that

$$\begin{aligned} V_c(x, y, z) &= V_c(-x, y, z) = V_c(x, -y, z) = V_c(x, y, -z) \\ &= V_c(y, x, z) = V_c(x, z, y) = V_c(z, y, x) \end{aligned}$$

Suppose now we taylor expand the potential about the origin as

$$\begin{aligned} V_c(\mathbf{r}) = V_c(\mathbf{0}) &+ a_1 x + b_1 y + c_1 z + a_2 x^2 + b_2 y^2 \\ &+ c_2 z^2 + a_{21} xy + b_{21} yz + c_{21} zx + \cdots \end{aligned} \tag{25}$$

with the terms $(a_1, b_1, \ldots)$ being related to the various derivatives of the potential at the origin and are constants. However, due to the O_h symmetry, not all terms are present. Moreover, several other nonvanishing terms are related to each other. Specifically, the linear terms are absent and $a_2 = b_2 = c_2 \equiv A$. Similarly, the cross term coefficients $a_{21} = b_{21} = c_{21} = 0$ due to the octahedral symmetry. At higher orders cross terms like $x^3 y$ cannot be present. Thus, the form of the potential which respects the O_h symmetry has to be

$$V_c(\mathbf{r}) = V_c(\mathbf{0}) + A(x^2+y^2+z^2) + B(x^4+y^4+z^4) + C(x^2y^2+y^2z^2+z^2x^2) + \cdots \tag{26}$$

The various constants A, B, and C involve the derivatives of the potential. Notice that the form above respects the symmetry relations in eq. (25), as desired. An important constraint on the potential has to do with the fact that it has to satisfy the Laplace equation i.e.,

$$\nabla^2 V_c(\mathbf{r}) \equiv \left(\frac{\partial^2}{\partial x^2} + \frac{\partial^2}{\partial y^2} + \frac{\partial^2}{\partial z^2} \right) V_c(\mathbf{r}) = 0 \tag{27}$$

This condition implies the following relation

$$6A + (12B + 4C)(x^2 + y^2 + z^2) = 0 \tag{28}$$

for any (x, y, z) which in turn leads to $A = 0$ and $C = -3B$. Using these relations we can now express the crystal field potential experienced by the electron due to the octahedral point charges as

$$V_c(\mathbf{r}) = V_c(\mathbf{0}) + \frac{5B}{2} \left[x^4 + y^4 + z^4 - \frac{3}{5} r^4 \right] + \cdots \tag{29}$$

The result above is identical, upto the fourth order, to the expression we derived earlier using electrostatics!

4. Molecular bonding and graph theory – a different kind of symmetry

As you might already know, symmetry principles play a crucial role in molecular structure and bonding. Group theory is an extremely powerful tool which lets us gain valuable insights into the molecular world. The utility is even greater when one would like to compute the molecular orbitals and the associated energies. In fact, modern computational chemistry packages use all the "bells and whistles" of group theory to their advantage. At the most elementary level, the symmetry of a molecule is essentially related to the way the various atoms are connected. If we were to discount certain extreme types of motion then the point group of the molecule is sufficient to glean important information about structure, bonding and spectra of the molecule. In this section I will use a particularly simple model for describing the π-bonding nature of conjugated molecules. This simple model [16] was proposed by Erich Armand Arthur Joseph Hückel in 1930 and has had an enormous influence[5] on the field of theoretical organic chemistry.

In applying Hückel theory to a molecule with N carbon atoms, one imagines a π-orbital, denoted by the wavefunction ϕ_k, centered on each of the carbon atom labeled by $k = 1, 2, \ldots, \mathrm{N}$. In this model one considers the overlap to be nonzero only between the π-orbitals centered on adjacent carbon atoms. Moreover the overlaps are the same between every pair of adjacent atoms and equal to unity i.e., one is assuming complete overlap of the orbitals. The π-molecular orbitals of the molecule can be viewed as a linear combination of the atomic orbitals

$$\psi_a = \sum_{k=1}^{N} c_{ak} \phi_k \tag{30}$$

The coefficients in the above expansion determine the extent of participation of the specific atomic orbitals. Thus, c_{ak} tells us about the extent of contribution of the atomic orbital ϕ_k centered on the kth-atom to the αth-molecular orbital. The theory provides a way to obtain the coefficients and hence determine the molecular orbitals as well as the energies ϵ_a associated with them. Note that one will have N molecular orbitals arising from the linear combination of the atomic orbitals on the N carbon atoms. There are atleast two perspectives on the use of group theory to obtain the molecular orbitals. One is a standard way wherein the point group of the molecule is used to determine the appropriate symmetry-adapted linear combinations. This can be found in almost every textbook on chemical applications of group theory. What I would like to present in these notes is a different, but related, perspective. I present it here since not many texts highlight the current approach which involves topology and graph theory. Moreover, this is an opportunity to illustrate the principles of symmetry from a different, and somewhat unusual, viewpoint. The description that follows is perhaps more algebraic than the usual geometric viewpoints espoused in standard texts. However, in this context it is intriguing to hear what the great chemist Robert Burns Woodward had to say [17].

"I remember very clearly – and it still suprises me somewhat – that the crucial flash of enlightenment came to me in algebraic, rather than in pictorial or geometric form. Out of the blue, it occured to me that the coefficients of the terminal terms in

[5]Incidentally, Hückel came up with his model during times of political turmoil and severe financial difficulties. Moreover, his work was not even appreciated by chemists until another two decades! This is all the more tragic when one realizes that his first paper in 1930 was on the stability of the benzene molecule due to the delocalization of the π-electrons. Remember, quantum mechanics appeared only in 1927 and people were still trying to grapple with the difficult issue of bonding in the hydrogen molecule.

the mathematical expression representing the highest occupied molecular orbital of butadiene were of opposite sign ..."

Note that this flash of enlightenment ultimately led to the famous Woodward–Hoffmann rules [18] which use orbital symmetry to predict the stereochemistry of pericyclic reactions. I say that the above statement is intriguing because Woodward, being the master of organic synthesis, would be expected to be more pictorial or geometric in his thinking. Of course, the algebraic approach has links to the geometric approach of point group theory in terms of the group operations on the molecule itself[6].

To begin let us first discuss the notion of alternant and non-alternant systems. Since we are interested only in the π-bonding nature, think of the molecule as a connected network of carbon atoms. More precisely, we think of the molecule as a graph with the vertices being the carbon atoms and the edges being the bonds that connect the atoms. It is crucial to note that in this way of representing the molecule one does not worry about the hydrogen atoms and all C–C bonds, single, double or triple, are considered equivalent. This latter assumption might look a bit strange but remember that the Hückel method, atleast the one that I am describing for you, is applicable to conjugated systems. As you all know very well, in conjugated π-bonded systems the π-electrons are delocalized and hence all C–C bonds are similar on the average. An additional fact that is ignored is the conformational aspect of the molecule. Thus, for instance, *cis*-ethylene and *trans*-ethylene are not distinguished and have the same connectivity.

Now, having the graph of a molecule in hand let us imagine placing a symbol (dot, star, etc.) on any carbon atom to start with and then subsequently placing the same symbol on alternating carbons. If this can be done without any problems then the molecule is called as an alternant system. Otherwise we call the molecule as non-alternant. So, naphthalene is an alternant system whereas fulvalene is a non-alternant system. The important point to note here is that the alternant systems represent a sort of symmetry. Indeed, they can be called as bichromatic systems since all the vertices can be labeled using only two colors. In Fig. 2 some examples of such bichromatic graphs are shown. The notion of bichromaticity allows us to distinguish between odd and even alternant molecules. In even alternant systems there are an equal number of the two colors used to label the graph. In contrast, for odd alternant systems one of the colors is in excess over the other. In Fig. 2, both butadiene and naphthalene are even alternant whereas the benzyl radical is odd alternant. How does this symmetry help us in understanding the π-bonding nature of the molecule? It so happens that several elegant results have been established in the field of graph theory and these results can be brought to bear on the molecules of interest [19]. I will not attempt to state or prove theorems in graph theory that have been very useful in the context of Hückel theory since several articles already exist in the literature. Here I will merely state one of the results and the interested reader can consult the literature for the relevant details.

---

[6]Remember that Evariste Galois' work on the impossibility of analytic solutions to a general quintic or fifth order equation is based entirely on group theoretic principles. In fact Galois is considered to be the first to develop group theory!

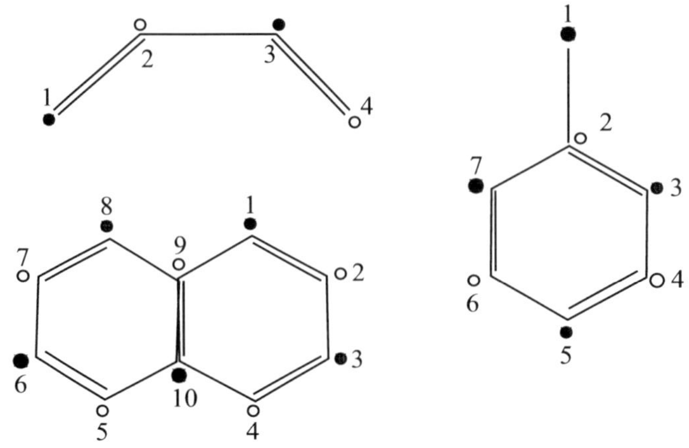

Figure 2. Illustrating bichromatic graphs. The molecules shown here are examples of alternant systems. Each carbon atom (vertex) can be alternatingly labeled by symbols (filled and open circles) with no ambiguity. The two molecules on the left column (butadiene and naphthalene) are examples of even alternant systems. The benzyl radical on the right is an example of an odd alternant system.

A particularly elegant result[7] pertaining to alternant (or bichromatic) systems is the Coulson-Rushbrooke pairing theorem [20]. According to the pairing theorem, for alternant systems

1. The π-electron energy levels come in pairs i.e., for every energy level $\alpha + x\beta$ there is another one of the form $\alpha - x\beta$. This immediately implies that for odd alternant systems there must be a single energy level with value α. In chemical terms this single energy level represents a nonbonding molecular orbital.
2. The molecular orbitals of the paired energy levels are closely related and differ only in the sign, not the magnitude, of the coefficients of the π-atomic orbitals on the various carbon atoms.
3. The total π-electron density is unity at every carbon atom.

Essentially the pairing theorem tells us that for alternant systems you get one molecular orbital and get another one for free. Let me illustrate this with the simple example of butadiene which has four π-electrons. In this case one finds that the energy of the highest occupied molecular orbital (HOMO) to be $\alpha - 0.618\beta$. Hence, butadiene being an even alternant molecule must have another energy level equal to $\alpha + 0.618\beta$, which happens to be the lowest unoccupied molecular orbital (LUMO). Now the HOMO turns out to be

$$\psi_{\text{HOMO}} = 0.60\phi_1 + 0.37\phi_2 - 0.37\phi_3 - 0.60\phi_4 \tag{31}$$

with the ϕ's denoting the atomic orbitals on the four carbon atoms of butadiene. The pairing theorem immediately yields the LUMO as

$$\psi_{\text{LUMO}} = 0.60\phi_1 - 0.37\phi_2 - 0.37\phi_3 + 0.60\phi_4 \tag{32}$$

[7] Interestingly, the analog of the Coulson-Rushbrooke theorem in pure mathematics appeared later. The famous Perron-Frobenius theorem on nonnegative, irreducible matrices was a harbinger to the pairing theorem.

Notice that the magnitude of the coefficients remain the same but some of them have opposite signs when compared to the expression for $\psi_{\rm HOMO}$. If you now look at the bichromatic graph for butadiene in Fig. 2 then it becomes clear that the signs have flipped only for the unstarred carbon atoms. This is a very general result and shows how the alternant symmetry of the graph yields the molecular orbital without any work. There are other examples in this context that one can provide, but I will not go into them here. Suffice it to say that the subject of algebraic graph theory provides rigorous connections between the eigenvalues (spectrum) of a graph and its symmetry properties.

Before finishing this section let me note another level of symmetry in the alternant systems. The pairing theorem above only requires the alternant character of the graph and does not need the bichromatic nature beyond this trivial level of symmetry. However, suppose we now take a bichromatic system and switch the two colors (or symbols) – is there a geometric operation which can switch back the colors and give us the original system? It is important to appreciate that this is a symmetry of the system which now explicitly involves the colors and hence systems which exhibit this symmetry can be called a bichromatically symmetric. The key point here is that not all alternant systems are bichromatically symmetric! Is this symmetry useful or is it just an interesting observation? Turns out that for bichromatically symmetric systems eigenvalues and eigenvectors can be obtained from a smaller system and hence the symmetry here provides for considerbale reduction in computational effort [21]. Let me end with a quote from the paper [18] of Hoffmann and Woodward.

"*In our work we have relied on the most basic ideas of molecular orbital theory – the concepts of symmetry, overlap, interactions, bonding, and the nodal structure of wave functions. The lack of numbers in our discussion is not a weakness – it is its greatest strength. Precise numerical values would have to result from some specific sequence of approximations. But an argument from first principles of symmetry, of necessity qualitative, is in fact much stronger than the deceptively authoritative numerical result. For, if the simple argument is true, then any approximate method, as well as the now inaccessible exact solution, must obey it.*"

5. Symmetry and dynamics

In the previous sections I have discussed examples which illustrate the use of symmetry arguments in the context of molecular structure and bonding. However, symmetry plays a crucial role in kinetics and dynamics of molecular reactions as well. The Woodward–Hofmann rules, alluded to earlier, rely upon the principle of conservation of orbital symmetry to decide whether a given reaction would be allowed or forbidden. This is one aspect of chemical reactions where symmetry decides the outcome. There are other aspects which arise mainly from the existence of identical atoms in the reaction species and one has to carefully keep track of the so called symmetry numbers for a proper understanding and calculation of the reaction rates. Let me start things by using a simple but nice example of the reaction

$$\rm H + HI \longrightarrow [H_2I]^{\ddagger} \longrightarrow H_2 + I \tag{33}$$

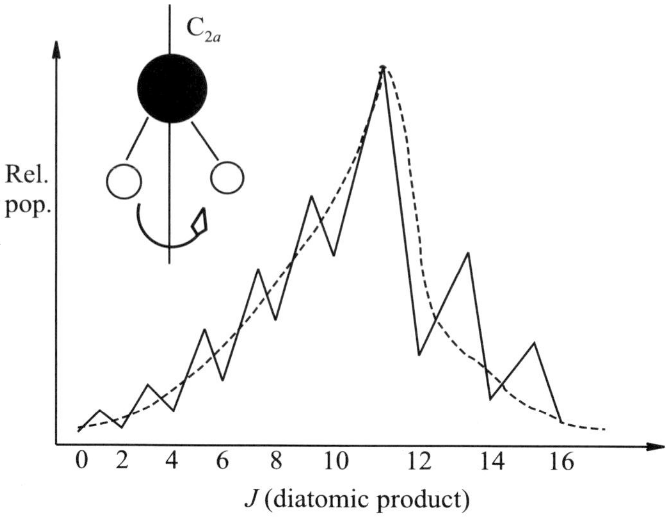

Figure 3. A sketch of the experimental results of the relative populations of the product hydrogen molecules in specific rotational state labeled by the rotational quantum number J. For the actual figure see the original work [22]. The alternating populations (solid line) for the even and odd J values is a direct consequence of the fermionic nature of the spin-1/2 hydrogen nuclei. The activated complex (shown) has C_{2v} symmetry and exchange of the hydrogens by a rotation must lead to a sign change in the overall wavefunction of the complex. Dividing the odd-J populations by three smoothes out the oscillations (dashed line).

which was studied in detail by Richard Zare and coworkers [22]. In the above reaction hydrogen atoms react with hydrogen iodide to yield hydrogen molecule and iodine atom as products. Experimetally, one can measure the relative populations of the rotational states of the hydrogen molecule[8]. The reaction, as indicated, goes through a transition state (activated complex) which has a bent geometry and corresponds to the point group symmetry C_{2v}. Such a symmetry means that if the two identical hydrogen atoms are swapped (via a C_2 rotation about the principal axis, for instance) then the transition state is left unchanged. When the experiment was performed, it was observed (see Fig. 3) that the odd rotational angular momentum states of H_2 were having higher populations as compared to the even rotational states. Moreover, as shown in Fig. 3, the data of the pouplation versus the rotational states showed an oscillatory behavior arising from the alternating high and low populations. One possible, and very exciting one at that, explanation involves interpreting the oscillations as evidence for reactive resonances associated with the activated complex – a much sought after effect in molecular beam studies. How does one understand the experimental observations?

[8]Experimetalists can now perform very detailed measurements. For instance, one can prepare the reactants in specific quantum states corresponding to particular rotational, vibrational, and electronic excitations and tell us the probability of observing the product hydrogen molecule in specific rovibrational states. Such experiments are called as crossed molecular beam experiments and have contributed enormously to our fundamental understanding of reaction dynamics and mechanisms.

The key point to be noted here is that the activated complex involves two identical hydrogen nuclei, which are spin-1/2 particles and hence belong to a class of particles called as fermions. In contrast, integer spin particles are called as bosons. Now, it turns out that exchanging two identical fermions or bosons has deep implications about the overall symmetry properties of the system. This symmetry is subtle since it happens at the level of the system wavefunction, and therefore not apparent from the geometric viewpoint. Suppose one has a system of two identical particles. The wavefunction of the system depends on the positions $\mathbf{r}_1$, $\mathbf{r}_2$ of both the particles. Then, mathematically, under exchange the wavefunctions behave as

$$\begin{aligned} \Psi_F(\mathbf{r}_1, \mathbf{r}_2) &= -\Psi_F(\mathbf{r}_2, \mathbf{r}_1) \\ \Psi_B(\mathbf{r}_1, \mathbf{r}_2) &= +\Psi_B(\mathbf{r}_2, \mathbf{r}_1) \end{aligned} \tag{34}$$

with F and B subscripts standing for a fermionic and bosonic system respectively. Hence, under exchange of the identical particles, the fermionic wavefunction is antisymmetric whereas the bosonic system is symmetric. Incidentally, this antisymmetry of the fermionic wavefunction is responsible for the Pauli exclusion principle that you are familiar with in the context of atomic structure. Since the hydrogen nuclei are fermions, the above symmetry implies that exchanging the hydrogens of the activated complex must result in a sign change of the overall wavefunction of $[H_2I]^{\ddagger}$. The overall wavefunction has electronic, vibrational, rotational, and nuclear spin parts of the complex. This means that if the complex is in a quantum state such that the electronic, vibrational, and rotational part of the wavefunction is symmetric under the exchange of the two hydrogens then, the nuclear spin part must be antisymmetric. This, in turn, implies that the H_2 product molecules will be exclusively found in antisymmetric rotational states (called as *para*). Conversely, the product hydrogen molecules will be in symmetric rotational states (called as *ortho*) if the electronic, vibrational, and rotational part of the wavefunction of the complex is antisymmetric under the exchange of the two hydrogen nuclei. What, if any, is the consequence of this symmetry on the reaction above? Under the reaction conditions, both odd and even rotational states of the complex are possible. Zare and coworkers looked [22] carefully at the data and realized that the odd states were nearly three times larger in population when compared to the even states. If the populations are weighted according to the *ortho-para* ratio of 3:1, then the oscillations go away and one has a smooth curve (see the dotted curve in Fig. 3). So, the experimental results did not indicate the presence of reactive resonances but were reflecting this deep symmetry of nature!

In fact people have worked out very general rules that govern the reaction dynamics. For example, Philip Pechukas showed [23] on very general grounds that the symmetry of a transition state can be no greater than the symmetry of the stationary points (think reactants and products) it connects. More specifically, symmetry of the transition state is identical to the symmetry that is exhibited by the system along its reaction path. Indeed, conservation of the symmetry along the reaction path implies that certain vibrational modes of the molecule cannot couple to the reaction coordinate [24]. In turn, this has important consequences for the dynamics of the reaction. For instance, pumping a lot of energy into a specific mode of a reactant molecule with the hope of increasing the rate might prove to be naive – that particular mode may be decoupled from the reaction coordinate or mode. Thus, the reaction rates may be quite different depending on which mode of a reactant molecule is initially excited and such effects are called as mode-specific effects. Therefore, knowledge of the dynamical effects of symmetry is essential for both understanding and manipulating a chemical reaction. In this context it is interesting to note that recent research in quantum control involves the notion of continuous symmetries, studied using the mathematical structure of Lie groups, to decide the controllability of a system [25]. The question of controllability can be stated as follows: *Given an external field that can be shaped at will, can a system be steered from one quantum state to another?* Answering this question is crucial to the field of optimal control theory and to the topology of the optimal control landscape.

5.1 *Nonrigid molecules: Role of permutation groups*

The example of fermions versus bosons above has another interpretation – essentially, we are permuting particles pairwise and looking at the effect on the overall wavefunction and dynamics of the system. It should not be very surprising that whenever one is dealing with a molecule with several identical atoms then the permutational symmetry must play a key role in unraveling the complex spectra and dynamics of the system. The concept of permutational symmetry has therefore been very well studied and is invoked to understand a wide range of chemical phenomena, from the structure of nuclear magnetic spectra to the spectra of weakly bound clusters [26]. Turns out that the general permutation symmetry, and the resulting permutational groups, contain the point groups as a special case[9]. In order to appreciate the need for the permutational symmetry, consider a molecule like PF_5 which exhibits fluxional behavior i.e., the fluorine atoms are exchanging amongst themselves even at low temperatures [27]. Clearly, attaching a point group symmetry for this molecule based on some rigid structure is not quite correct. There are large displacements in the PF_5 molecule and the system explores several different point group structures during its fluxional dynamics (known as Berry pseudorotation). A similar effect happens in the CH_5^+ molecule, for which researchers are just starting to gain insights into the structure and dynamics [28]. In general, for any molecule which exhibits large amplitude motions, the notion of a point group is of limited utility and one needs to invoke a larger set of symmetries to account for the spectrum as well as the dynamics.

In order to appreciate the need for a larger set of symmetries let us take[10] the example of methyl halide CH_3X. In terms of the usual point group symmetry considerations, this molecule belongs to the C_{3v} symmetry group. However, this

[9]It might interest you to know that permutational groups play the central role in Evariste Galois' theory about the solvability of algebraic equations. Furthermore, as shown by Felix Klein in an influential work, there are very elegant connections between quintic equations, icosahedral groups, permutations and elliptic functions. I must also mention the name of Niels Hendrik Abel in this context.

[10]The examples chosen here and the presentation rely heavily on the book by Bunker and Jensen [26]. This book is a prerequisite for anyone wishing to undertake a serious study of the nonrigid molecular spectra.

symmetry assumes a rigid equilibrium structure and in particular does not account for a possible inversion motion of the methyl protons. This is very much the same sort of motion, called as the umbrella inversion, in ammonia NH_3 that you are familiar with. If the molecule did undergo such inversion tunneling process then the fingerprints of this can be found in a spectroscopic experiment with sufficiently high resolution. One typical fingerprint would be the observation of closely spaced doublets (or multiplets) instead of a single line in the spectrum. In this case, using the C_{3v} point group of the molecule to understand the spectrum would not be of much help. What we would need at this stage would be a more general symmetry group – something bigger than the point group. How can we obtain such a group? In what follows, I will give you a brief description of the construction of such a group which is called as the complete nuclear permutation inversion (or CNPI) group. The point group C_{3v} will turn out to be a subgroup of the CNPI group.

To begin, let us label the methyl protons by the numbers $(1, 2, 3)$ as shown in the figure. Let us now consider the set of all possible permutations of the three hydrogens i.e., all the permutations of the three numbers. From your basic knowledge of permutation and combinations, it is clear that there are a total of $3! = 6$ possibilities. For example, one possible permutation is switching 1 with 2 with 3 remaining fixed. Mathematically, such an operation can be represented as

$$\begin{pmatrix} 1 & 2 & 3 \\ 2 & 1 & 3 \end{pmatrix} \equiv (12) \tag{35}$$

where we have used a shorthand notation (12) for the permutation operation. As another example consider cyclic permutation of all three numbers which can be represented as

$$\begin{pmatrix} 1 & 2 & 3 \\ 2 & 3 & 1 \end{pmatrix} \equiv (123) \tag{36}$$

Such a cyclic permutation is called as a *cycle*. Thus, one can construct a group consisting of all possible permutations of the three numbers amongst themselves and an additional element E which is the identity or 'do nothing' operation. This group is called as the complete nuclear permutation (CNP) group and denoted as

$$S_3 = \{E, (12), (23), (13), (123), (132)\} \tag{37}$$

To the above permutation operations we have to also add the operation, denoted by E^*, wherein one inverts the spatial coordinates of all the particles in the molecule – electrons and nuclei. This inversion operation is crucial if one wants to determine whether the molecular wavefunction has a definite parity or not[11]. How does one include the inversion operation? The CNP is already a group. So, in order to enlarge the CNP group we have to take a direct product with another group that includes the inversion. Since two inversions will bring the molecule back to its original

[11] Such a procedure is familiar to you from basic molecular orbital theory. For example, the molecular orbitals of homonuclear diatomics like H_2, N_2 etc. carry the *g* (gerade) or *u* (ungerade) labels. Thus, the notation $1\sigma_g$ or $1\pi_u$ reflect the inversion symmetry of the molecular orbitals.

form we can mathematically say that $E^*E^* = E$, the identity operation. Thus, $\{E, E^*\}$ is a group (known as the inversion group) and we can now enlarge the CNP group as

$$\{E, (12), (23), (13), (123), (132)\} \otimes \{E, E^*\} \tag{38}$$

and the above group is known as the complete nuclear permutation inversion or CNPI group of the molecule. Explicitly, by expanding out the direct product, one can write

$$G_{\text{CNPI}} = \{E, (12), (23), (13), (123), (132), E^*, (12)^*, (23)^*, (13)^*, (123)^*, (132)^*\} \tag{39}$$

and hence the CNPI group for a molecule like CH_3X has an order equal to twelve. Notice that the rigid C_{3v} point group is of order six. Hence the G_{CNPI} is twice the size of the point group. The key point here to note is that the CNPI group encompasses all of the potential large amplitude motions of the methyl hydrogens.

Once one has the CNPI group in hand, one has to think carefully about the possible dynamical motions of the atoms which will allow the molecule to explore several strcutures with distinct point group symmetries. For example, in the case of ammonia NH_3 whose usual pyramidal structure has C_{3v} symmetry, the umbrella inversion dynamics will include a configuration wherein the molecule is planar with D_{3h} symmetry. For more complicated systems, the molecule's rich and complex dynamics can explore one or more of the several degenerate to near-degenerate minima. Let us come back to the methyl halide example and understand when the CNPI group might prove useful. There are two ways of numbering the methyl hydrogens as shown in Fig. 4, with one version being clockwise (C) and the other being anticlockwise (A) when viewed along the C–X direction. Apart from the different numbering the two versions are physically the same and hence on the global potential energy surface of the molecule the A and C versions will correspond to symmetric minima. A highly simplified sketch[12] of the potential is also shown in Fig. 4, which also schematically indicates some of the energy levels. Now, note that a permutation like (12), which is an element of the G_{CNPI}, will interconvert between the A and C versions. Such an interconversion corresponds to the phenomenon of quantum mechanical tunneling from one minimum to the other. From the general principles of quantum mechanics, whenever tunneling is possible, the vibrational energy levels will be split as shown in the figure. However, the extent of splitting depends on the height, and thickness, of the barrier separating the two minima[13]. Specifically, if the barrier is infinitely high then there is no tunneling and the energy levels will be degenerate. On the other hand, if the barrier is finite then tunneling is possible and one can write down approximate expressions for the splitting using basic quantum mechanics. The significant point in

[12] Remember, since this is a five atom system, the dimensionality of the global potential energy surface will be equal to 9. So, the figure being one dimensional is an oversimplification. However, it is sufficient for our qualitative arguments.

[13] In multidimensions there are other factors to consider and the splitting can display a bewildering range of complexity. Nevertheless, in this simplified picture and description we will ignore this interesting and difficult aspect of the tunneling phenomenon.

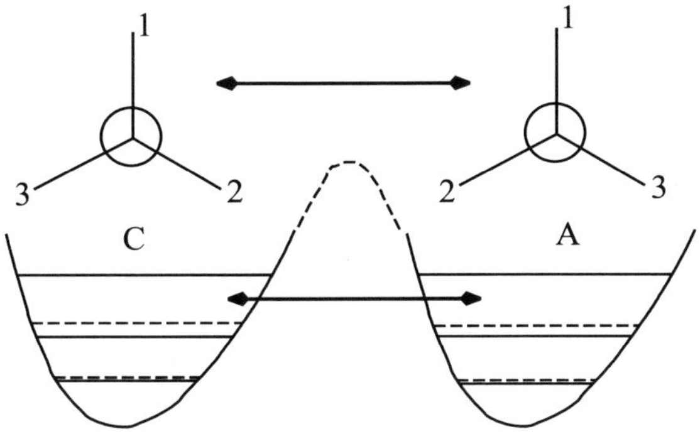

Figure 4. A sketch illustrating the two possible numberings of the methyl hydrogens in CH_3X. Note that the left version is denoted C for clockwise and the right version is denoted A for anticlockwise. Rotational motion about the C–X bond will not interconvert between the two versions. Since the two representations of the molecule are physically the same, they can be associated with two symmetric minima on the potential energy surface. A caricature of such a symmetric double minimum potential is also shown with some of the energy levels. The dotted part of the potential is drawn to suggest that the barrier might or might not be finite. A finite barrier will result in quantum tunneling between the A and C versions leading to a doublet structure (dotted line energy levels) of the spectrum.

this context is that a sufficiently high resolution spectroscopic experiment will reveal the splittings, if any. If there is an observable splitting in the spectrum of CH_3X then one absolutely needs the G_{CNPI} to identify, label, and hence assign the spectrum. As it turns out, there does not seem to be any evidence for such inversion tunneling splittings in the spectrum of CH_3X. In that case, we have to remove those elements from the G_{CNPI} which interconvert between the A and C versions. What are these elements? As mentioned above, (12) is one such element and you can discover the other 'offending' elements as well. Upon removal of such group elements we are left with a smaller group

$$C_{3v}(M) = \{E, (123), (132), (12)^*, (23)^*, (13)^*\} \tag{40}$$

which is the *molecular symmetry group* of the molecule when the tunneling splittings are not observed. In contrast, for ammonia the splitting is observed [29] spectroscopically and hence an unambiguous assignment of the spectrum is only possible if the molecular symmetry group is taken to be the full CNPI group in Eq. 39. Incidentally, the splitting of the ground vibrational level in NH_3 is about $0.8\,\mathrm{cm}^{-1}$ and this means that the molecule is undergoing about 10^{10} inversions every second! In general, if we denote the splitting by Δ (in wavenumbers) then the time scale for tunneling is given by $\tau \sim 1/c\Delta$ with c being the speed of light.

The above, admittedly simple, example was chosen to introduce the notion and utility of a CNPI group. The real power of such an approach becomes apparent when one is faced with the complex spectral features that are observed for weakly bound clusters. For example, a proper characterization of the structure and dynamics of water clusters $(H_2O)_n$, as manifested in their spectra, requires one to invoke various aspects of the CNPI group. Specificallly, for the water dimer $(H_2O)_2$ the possible permutations involve four hydrogens and two oxygens and hence the CNPI group is mathematically

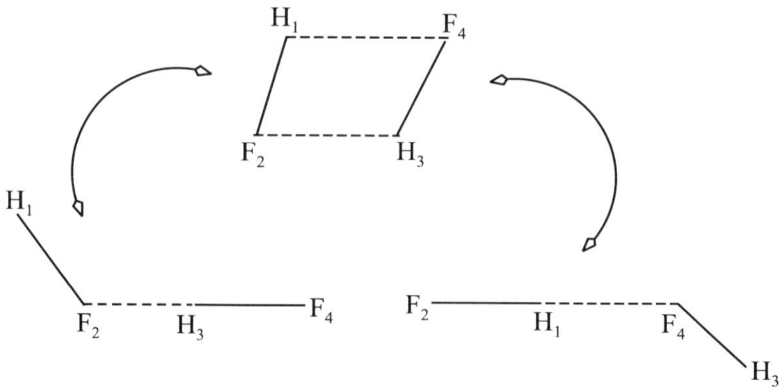

Figure 5. A sketch illustrating the two possible versions of the hydrogen fluoride dimer $(HF)_2$. Note that the left version and the right version can interconvert if the hydrogen bond is allowed to break and reform. Experimentally a splitting is observed and the figure shows a possible way in which the two versions can intercovert. The symmetry group G_4 is required to account for the spectral features associated with such a large amplitude motion. The pathway shown here corresponds to a *trans* pathway. There is also a *cis*-pathway which is not shown here.

represented as $S_4^{(H)} \otimes S_2^{(O)} \otimes \{E, E^*\}$ which has 96 elements. However, one needs to only use a subgroup of order 16 denoted by G_{16} in order to understand the vibration-rotation tunneling rearrangements that occur in the dimer [30].

Another example comes from the hydrogen fluoride dimer $(HF)_2$ whose CNPI group is $S_2^{(H)} \otimes S_2^{(F)} \otimes \{E, E^*\}$ with a total of eight elements. Microwave spectroscopy by Dyke, Howard and Klemperer revealed [31] a splitting of about $0.65\,\text{cm}^{-1}$ which involved tunneling, due to internal rotation, between two versions of the dimer by breaking and reforming hydrogen bonds. This is illustrated in Fig. 5 with two different versions of the dimer and their interconversion. Notice that an operation like (13) transfroms the system to a version wherein the strong H–F bond needs to be broken. So, we have to remove this from the G_{CNPI}. However, the operation (13)(24) is fine since, as shown in Fig. 5, this corresponds precisely to the tunneling motion. You can easily check that in this case the required molecular symmetry group is the four element group

$$G_4 = \{E, E^*, (13)(24), (13)(24)^*\} \tag{41}$$

On the other hand, if the tunneling splitting was not observed at all in the experiments then the molecule has $C_s(M) = \{E, E^*\}$ symmetry since the molecule is planar and there is a horizontal plane of symmetry. Incidentally, the $(HCl)_2$ dimer is much more floppy in comparison with a splitting of about $15.5\,\text{cm}^{-1}$.

The few simple examples above were chosen to introduce you to the subject of large amplitude motions, floppy molecules and the power of symmetry arguments to understand the spectra and dynamics of such systems. Suffice it to say that this is just the tip of the iceberg and the topic of large amplitide motions and tunneling is at the forefront of current research. Remember, the structure of a molecule only becomes clear when we can understand the spectrum and hence one cannot do without the principles of symmetry.

6. Conclusions

I hope that this lecture has given a glimpse of the use of symmetry and the language of group theory as applied to structure and dynamics of molecules. Several excellent texts have been written which will provide further examples which utilize the full power of group theory. In essence, what I have shown is that group theory lets us understand the patterns that are observed in molecular spectra. Towards the end, however, I have tried to convey the central message that structure and dynamics are intricately connected, perhaps even entangled. This interdependence of structure and dynamics is presumably what Hoffmann meant by the words "tense edge of symmetry", since the feasibility of specific pathways which interconvert between different forms of the molecule determines how symmetric (or asymmetric) a system is. It is appropriate to end with a quote from the book [32] by Lefebvre-Brion and Field "*Herzberg taught us about the patterns that are likely to be found in a spectrum, how to use these patterns to make secure assignments, and how to build models that reproduce the measurements to the precision of the experiment. But patterns are made to be broken, and the breaking of standard patterns is the key to perceiving those dynamical features that demand explanation. No spectrum, no matter how simple, is dynamics-free.*" Group theory lets us recognize the patterns and when they are broken it is group theory again which tells us how the patterns are broken!

References

[1] H. Weyl, Symmetry, Princeton university press, New Jersey (1952).

[2] B. C. Berndt and R. A. Rankin, Ramanujan: Letters and commentary, History of mathematics, volume 9, American Mathematical Society, Providence (1995).

[3] B. Mandelbrot, The fractal geometry of nature, W. H. Freeman & Co, San Francisco (1982).

[4] L. Lam, Nonlinear physics for beginners: Fractals, chaos, solitons, pattern formation, cellular automata and complex systems, World Scientific, NJ (1998); H.-O. Peitgen and P. H. Richter, The Beauty of Fractals, Springer-Verlag, Berlin (1986).

[5] R. Hoffmann, *Angew. Chem. Int. Ed. Engl.*, **32** (1993) 129.

[6] E. Heilbronner and J. D. Dunitz, Reflections on symmetry in chemistry ... and elsewhere, VHCA, Verlag Helvetica Chimica Acta, VCH verlagsgesellschaft mbH, Weinheim (1993).

[7] E. Merzbacher, The early history of quantum tunneling, *Phyics Today*, **55** (2002) 44.

[8] There are several books on symmetry applications in chemistry. For example see, F. A. Cotton, Chemical applications of group theory, Third edition, Wiley India Pvt Ltd, New Delhi (2009); D. M. Bishop, Group theory and chemistry, Dover publications (1993); I. Hargittai and M. Hargittai, Symmetry through the eyes of a chemist, Third edition, Springer (2009).

[9] Nobel lectures, Chemistry 1996–2000, Edited by I. Grenthe, page 89, World Scientific, Singapore (2003).

[10] Nobel lectures, Chemistry 1996–2000, Edited by I. Grenthe, page 44, World Scientific, Singapore (2003).

[11] M. Livio, The golden ratio: The story of phi, the world's most astonishing number, Broadway books, NY, 2002; H. Walser, The golden section, Mathematical Association of America (2001).

[12] G. H. Hardy and E. M. Wright, An introduction to the theory of numbers, fifth edition, Clarendon press, Oxford (1979).

[13] F. Rioux, Quantum mechanics, group theory, and C_{60}, *J. Chem. Ed.*, **71** (1994) 464; A. F. Hebard, Superconductivity in doped fullerenes, *Physics Today*, **45** (2002) 26.

[14] H. A. Bethe, Splitting of terms in crystals, *Ann. Phys.*, **3** (1929) 133; See also the article by H. R. Krishnamurthy, Bethe's contribution to solid state theory: A tribute, *Resonance*, **10** (2005) 55.

[15] J. H. Van Vleck, Theory of the variations in paramagnetic anisotropy among different salts of the iron group, *Phys. Rev.*, **41** (1932) 208.

[16] E. Hückel, Zur quantentheorie der doppelbindung (On the quantum theory of double bonding), *Zeit. für Physik*, **60** (1930) 423; See also, J. Chandrasekhar, Erich Hückel (1896–1980), *Resonance*, **3** (1998); H. Hartmann and H. C. Longuet-Higgins, Erich Hückel, *Biographical Memoirs of Fellows of the Royal Society*, **28** (1982) 153; J. A. Berson, Erich Hückel, Pioneer of organic quantum chemistry: Reflections on theory and experiment, *Angew. Chem. Intl. Ed. Engl.*, **35** (1996) 2750.

[17] O. T. Benfey and P. J. T. Morris, Robert Burns Woodward: Architect and artist in the world of molecules, Chemical Heritage Foundation, Philadelphia (2001) 415; See also, R. Hoffmann, A claim on the developement of the frontier orbital explanation of electrocyclic reactions, *Angew. Chem. Intl. Ed.*, **43** (2004) 6586.

[18] R. Hoffmann and R. B. Woodward, Conservation of orbital symmetry, *Acc. Chem. Res.*, **1** (1968) 17.

[19] K. Balasubramanian, Applications of combinatorics and graph theory to spectroscopy and quantum chemistry, *Chem. Rev.*, **85** (1985) 599; N. Trinajstic, Chemical Graph Theory, CRC Press, Boca Raton (1992).

[20] C. A. Coulson and G. S. Rushbrooke, Note on the method of molecular orbitals, *Proc. Cambridge Phil. Soc.*, **36** (1940) 193.

[21] See for example, S. S. D'Amato, Eigenvalues of graphs with threefold symmetry, *Theor. chim. Acta (Berl.)*, **53** (1979) 319.

[22] D. A. V. Kliner, K. D. Rinnen, and R. N. Zare, Effect of indistinguishable nuclei on product rotational distributions: The $H+HI \to H_2+I$ reaction, *J. Chem. Phys.*, **90** (1989) 4625.

[23] P. Pechukas, On simple saddle points of a potential surface, the conservation of nuclear symmetry along paths of steepest descent, and the symmetry of transition states, *J. Chem. Phys.*, **64** (1976) 1516.

[24] W. H. Miller, Dynamical effects of symmetry along a reaction path: Mode specificity in the unimolecular dissociation of formaldehyde, *J. Am. Chem. Soc.*, **105** (1983) 216.

[25] G. M. Huang, T. J. Tarn, and J. W. Clark, On the controllability of quantum-mechanical systems, *J. Math. Phys.*, **24** (1983) 2608; S. G. Schirmer, H. Fu, and A. I. Solomon, Complete controllability of quantum systems, *Phys. Rev. A*, **63** (2001) 063410; D. D'Alessandro, Introduction to Quantum Control and Dynamics, CRC Press, Boca Raton, Florida (2007).

[26] P. R. Bunker and P. Jensen, Molecular Symmetry and Spectroscopy, Second edition, NRC Research Press, Ottawa, Ontario, Canada (1998).

[27] M. E. Cass, K. K. Hii, and H. S. Rezpa, Mechanisms that interchange axial and equatorial atoms in fluxional processes: illustration of the Berry psuedorotation, the turnstile, and the lever mechanisms via animation of transition state normal vibrational modes, *J. Chem. Ed.*, **83** (2006) 336.

[28] E. T. White, J. Tang, and T. Oka, CH_5^+: The infrared spectrum observed, *Science*, **284** (1999) 135; O. Asvany, P. P. Kumar, B. Redlich, I. Hagemann, S. Schlemmer, and D. Marx, Understanding the infrared spectrum of bare CH_5^+, *Science*, **309** (2005) 1219; X. Huang, A. B. McCoy, J. M. Bowman, L. M. Johnson, C. Savage, F. Dong, and

D. J. Nesbitt, Quantum deconstruction of the infrared spectrum of CH_5^+, *Science*, **31** (2006) 60.

[29] P. J. Gordon, J. H. Zeiger, and C. H. Townes, Molecular microwave oscillator and new hyperfine structure in the microwave spectrum of NH_3, *Phys. Rev.*, **95** (1954) 282.

[30] K. Liu, J. D. Cruzan, and R. J. Saykally, Water clusters, *Science*, **271** (1996) 929; D. J. Wales, Rearrangements and tunneling splittings of protonated water dimer, *J. Chem. Phys.*, **110** (1999) 10403; K. Balasubramanian and T. R. Dyke, Nuclear spin species, statistical weights, and correlation tables for weakly bound van der Waals complexes, *J. Phys. Chem.*, **88** (1984) 4692.

[31] T. R. Dyke, B. J. Howard, and W. Klemperer, Radiofrequency and microwave spectrum of the hydrogen fluoride dimer; a nonrigid molecule, *J. Chem. Phys.*, **56** (1972) 2442.

[32] H. Lefebvre-Brion and R. W. Field, The Spectra and Dynamics of Diatomic Molecules, Elsevier, Amsterdam (2004).

Symmetry: A Multi-Disciplinary Perspective No. 16, 2011, pp. 97–117.

Group Theory and Tiling Problems

B. Sury

Stat-Math. Unit, Indian Statistical Institute, 8th Mile Mysore Road, Bangalore 560 059, India
e-mail: sury@isibang.ac.in; surybang@gmail.com

Abstract. Tiling subsets of the plane by polygonal shapes or more generally of subsets of higher-dimensional spaces by polyhedra is intimately connected with group theory. The factorization theory of groups plays a key role and deeper aspects are studied with the aid of the so-called tiling group. In this article, we give a sampling of some tiling problems including the classical Minkowski conjecture and it solution using group theory. The discussion is far from exhaustive and merely scratches the surface of an intimate friendship of group theory and geometry through a study of how they work together in unravelling some secrets of symmetry.

Keywords. Minkowski conjecture, tiling group, factorization theory of groups, tiling and music.

To find if it is possible to tile
or if any attempt has to be futile,
group theory can go many a mile
and serve symmetry in style.
Groups and symmetry –
a friendship of algebra and geometry

0. Introduction

This article is an extended version of a talk delivered in the 'Symposium on symmetry' at IISER Mohali in February 2010[1]. It will feature some elementary problems of geometry and combinatorics where the main tour-de-force is group theory. The symposium featured wonderful talks by physicists, biologists, chemists and even architects which shows the ubiquity of the theme of symmetry. The magic of this subject was expressed during the inauguration of the symposium on symmetry by my friend Kapil Paranjape who called it *"Khul ja sym-sym"*! I talk about one part of the aspect of symmetry which is elementary on the face of it, viz., tiling, and indicate the role of group theory. Everyone has seen the tiling of the plane by equilateral triangles, or by squares or by regular hexagons. These are examples of lattice tilings of the plane. One says that a set of lattice figures tiles a region R if the region can be covered by the translates of the tiles without intersections. If a region R is specified and a finite set of tiles are specified, the question whether R admits a tiling by these tiles is a decidable problem which is NP-complete. It is unknown whether there is a polynomial time

[1] In admiration of Mrs. S. Devi–a teacher par excellence-who passed away on 8th February 2010, 10 days before this symposium

algorithm to solve the finite tiling problem but one is unlikely to exist. More generally, one could look at tiles of different shapes and, one could look at tiling problems in $\mathbf{R}^n$. Many such problems can be naturally rephrased (and solved!) using group theory. This is a very exciting and alive area of mathematics.

Regarding quoting references, there are far too many papers on this subject over the past 100 years to even mention a reasonable fraction of them. We give only some of the more modern ones which we have referred to and which would be easy for the interested reader to get hold of. We particularly point out to the books [22] and [24]. The latter book is referred to in section 4 and the former is used in section 2 predominantly. Section 3 refers to the wonderful paper by Conway & Lagarias ([3]); [19], [20] are two other important papers with newer results. References [1], [2] and [13] can be consulted for more details on section 5. The old papers by Keller [12], Minkowski [16] and Perron [21] are mentioned in section 1 where [15] can be consulted for more details. The paper by Hajós [7] can be consulted for the original proof of Minkowski's conjecture detailed in section 4. All other references are related to section 2.

1. Minkowski's and Keller's conjectures

A classical conjecture in 1907 due to Minkowski ([16], see §4 of chapter 2 and §7 of chapter 3 – the conjecture appears on p. 28 and its geometric interpretation on p. 74.) asserts:

Conjecture (Minkowski). Every lattice tiling of $\mathbf{R}^n$ by unit cubes contains two cubes that intersect in an $(n-1)$-dimensional face.

Informally, a tiling is a partition of some infinite space (usually $\mathbf{R}^n$) into pieces having a finite number of distinct shapes. More precisely, a subset T of $\mathbf{R}^n$ is said to tile it by translations if there is a discrete subset Λ of $\mathbf{R}^n$ (the translation set) such that, up to sets of measure zero, the sets $T+\lambda(\lambda \in \Lambda)$, are disjoint and their union is the whole of $\mathbf{R}^n$. One writes $\mathbf{R}^n = T \oplus \Lambda$. One talks of lattice tilings when Λ a lattice. Roughly, a lattice is an infinite arrangement of points spaced with sufficient regularity that one can shift any point onto any other point by some symmetry of the arrangement. Let us define all these more precisely now. Formally, a lattice Λ in $\mathbf{R}^n$ is the set of all integer linear combinations of a vector space basis of $\mathbf{R}^n$. The simplest and most common example of a lattice is formed by the points all coordinates of which are integers. For example, if T is a single unit square in $\mathbf{R}^2$, then it gives a lattice tiling of the plane for the lattice $\mathbf{Z}(1,0)+\mathbf{Z}(0,1)$. The lattice $\Lambda = \mathbf{Z}(1,0)+\mathbf{Z}(1/2,\sqrt{3}/2)$ satisfies $\Lambda \oplus R = \mathbf{R}^2$ where R is the parallelogram made of the two equilateral triangles in the figure below – its vertices are at $(0,0)$, $(1/2,\sqrt{3}/2)$, $(1,0)$, $(3/2,\sqrt{3}/2)$. We mention in passing that there are also sets which tile $\mathbf{R}^n$ by some translation sets, but which do not admit lattice tilings. For instance, the union $[0,1/2] \cup [1,3/2]$ tiles $\mathbf{R}$ with the translation set $\Lambda = \{0,1/2\}+2\mathbf{Z}$, and yet we do not have a lattice tiling.

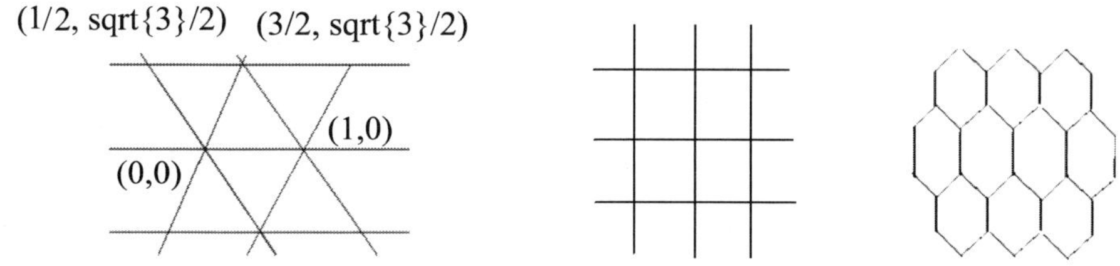

(a) Triangular (b) Square (c) Hexagonal

Group-theoretic reformulation of Minkowski's conjecture due to Hajós. Consider a possible counter-example to the conjecture; that is, there is a lattice Λ and a unit cube T such that $\mathbf{R}^n = \Lambda \oplus T$ but no two cubes $\lambda + T$, $\mu + T$ have a full common $(n-1)$-dimensional face. If some co-ordinates of some basis vectors of Λ are irrational, we may perturb all the co-ordinates slightly to make them rational in such a way that we still retain this as a counter-example to the conjecture. This latter condition can be ensured as follows. It suffices to ensure that no cube meets the cube centred at the origin in an $(n-1)$-dimensional face. As only finitely many cubes touch this central cube, a sufficiently small perturbation would indeed preserve the property of being a counter-example. Thus, we now have a counter-example where all the vectors of the basis of Λ have rational entries. Thus, we can choose an integer r (for example, the least common denominator of all the entries of the basis vectors) such that all our cubes are centred at points whose entries are integral multiples of $1/r$. If $\Lambda' := \oplus_{i=1}^{n} \mathbf{Z}\frac{e_i}{r}$, then Λ is a sublattice of it. The quotient group Λ'/Λ is a finite abelian group. Instead of placing the cubes centred at the points of Λ (which was our tiling), let us place them with one corner at a point of Λ (the corner with the smallest entry in each co-ordinate dimension-this is the lower left corner for $n = 2$). As each point of $\mathbf{R}^n$ is uniquely obtained on the translation of a unit cube by the elements of Λ, the elements of Λ'/Λ correspond to lattice points in one of these unit cubes. So, each element of Λ'/Λ can be expressed as $\sum_{i=1}^{n} x_i \frac{e_i}{r}$ with $0 \leq x_i < r$. Two of the cubes in the tiling will meet in an $(n-1)$-dimensional face if, and only if, for some $i \leq n$, we have $e_i \in \Lambda$; that is, $r\frac{e_i}{r} \in \Lambda$.

In other words, the reformulation of Minkowski's conjecture in terms of the group Λ'/Λ is:

Conjecture of Minkowski as reformulated by Hajós in 1938. Let G be an abelian group, r a positive integer and $a_1, \ldots, a_n \in G$ be such that each element of G is expressible uniquely as $x_1 a_1 + \cdots + x_n a_n$ with $0 \leq x_i < r$. Then one of the a_i's must satisfy $r a_i = 0$.

In fact, Hajós made the following (apparently) more general assertion which he proved in 1941 thereby settling Minkowski's conjecture:

Let G be an abelian group, $r_1, \ldots, r_n$ be positive integers and $a_1, \ldots, a_n \in G$ be such that each element of G is expressible uniquely as $x_1 a_1 + \cdots + x_n a_n$ with $0 \leq x_i < r_i$. Then one of the a_i's must satisfy $r_i a_i = 0$.

As a matter of fact, it is a nice exercise to show that this general version can be deduced from the special version above which arises in Minkowski's conjecture but we do not go into that here! But, let us try to understand the group theory better so that we may later discuss the solution.

In a finite, abelian group G (written additively as before), look at an element g and a set of the form $\{0, g, 2g, \ldots, rg\}$. If r is less than the order of g, these distinct elements form the front end of the cyclic group generated by g. Such a set is called a *cyclic subset*. Note that if $r + 1 = O(g)$, then this cyclic subset is actually a subgroup. If $A_1, \ldots, A_n$ are subsets of G such that each element of G is expressible uniquely in the form $a_1 + a_2 + \cdots + a_n$ with $a_i \in A_i$, one calls this as a *factorization* of G by the subsets $A_1, \ldots, A_n$. Now then, Hajós's statement is simply:

In any factorization of a finite abelian group by cyclic subsets, at least one of the factors is a subgroup!

Thus, this is the dual statement to the so-called fundamental theorem of finite abelian groups which asserts that any finite abelian group admits a factorization by cyclic subgroups. We shall discuss the proof of Minkowski's conjecture in section 4.

In 1930, Keller conjectured that Minkowski's conjecture ought to hold even when the tiling by unit cubes is not a lattice tiling. Keller's conjecture was verified by Perron in 1940 up to dimension 6. However, J. Lagarias and P. Shor showed in 1992 using a graph-theoretic reformulation of finding maximal cliques that Keller's conjecture is false for all dimensions from 10 upwards. Later, in 2000, John Mackey found a counter-example in dimension 8; it is still open only in dimension 7. There are other generalizations of Minkowski's conjecture due to Furtwangler etc. which too turned out to be false in general.

2. Tiling with rectangles and triangles and number theory

Moving from the tiling of whole Euclidean space, we talk about some finite tiling and covering problems which lead to interesting methods from number theory.

2.1 *Tling squares and polyominos by triangles*

Try to cut a square into finitely many triangles (possibly of different shapes) of equal area. You would find that – no matter what the shapes are – the number of triangles is

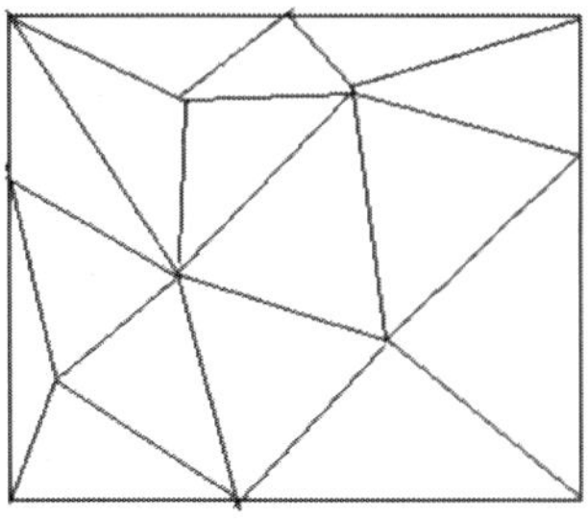

always even. Amazingly, the natural proof of this fact uses 2-adic valuations! In fact, if an n-dimensional cube is cut into simplices of equal volumes, it turns out that the number of simplices must be a multiple of $n!$ (!) The first assertion was proved by P. Monsky in 1970 and the second one was proved in 1979 by D. G. Mead.

Theorem (P. Monsky). *A subdivision of a square into finitely many triangles of equal areas must have an even number of triangles.*

There is also a generalization of Monsky's result due to S. K. Stein in 1999. This involves the so-called polyominos which are connected unions of unit squares. The most commonly occurring ones are the dominos which are unions of two unit squares. Special polyominos are those which have standard edges; i.e., edges of the unit squares are parallel to the co-ordinate axes. Stein's theorem asserts:

Theorem (S. K. Stein). *Consider a special polyomino which is the union of an odd number of unit squares. If this polyomino is a union of triangles of equal areas, then the number of triangles is even.*

Towards the proof of Monsky's and Stein's results, consider the "2-adic valuation map" $\phi : \mathbf{Q}^* \to \mathbf{Q}$ defined by $\phi(\frac{2^a b}{c}) = a$ where b, c are odd; define also $\phi(0) = \infty$.
Colour a point $(x, y) \in \mathbf{Q} \times \mathbf{Q}$ by the colour:

$$\begin{aligned} &\text{red}, && \text{if } \phi(x) > 0,\ \phi(y) > 0, \\ &\text{blue}, && \text{if } \phi(x) \le 0,\ \phi(x) \le \phi(y), \\ &\text{green}, && \text{if } \phi(x) > \phi(y) \text{ and } \phi(y) \le 0. \end{aligned}$$

In this manner, all points of $\mathbf{Q} \times \mathbf{Q}$ are coloured by these three colours. For example, $(2, 0)$ is red, $(1, 3)$ is blue and $(1, 1/2)$ is green. Also, $(0, 0)$ is red while $(1, 0)$ is blue. It is possible to extend this function ϕ to fields containing $\mathbf{Q}$. For finite extensions, it is easy but for $\mathbf{R}$, it is more involved. However, there does exist a function $\phi : \mathbf{R} \to \mathbf{R}$ which satisfies:

$$\begin{aligned} &\phi \text{ restricts to the 2-adic valuation on } \mathbf{Q}; \\ &\phi(xy) = \phi(x) + \phi(y); \\ &\phi(x + y) \ge \min(\phi(x), \phi(y)). \end{aligned}$$

For simplicity, let us discuss first Monsky's result itself where the polyomino is the unit square with a left lower corner at $(0, 0)$. Here are a few easy observations:

Lemma.

(i) *If a point a is red, then any point x and $x + a$ have the same colour.*
(ii) *On any line, there are at the most two colours.*
(iii) *The boundary of the square has an odd number of segments which have a red end and a blue end.*
(iv) *If a triangle is not 'complete' (that is, has vertices only of one or two colours), then it has 0 or 2 red-blue edges.*

Proof. Let us write $|x|_2 = 2^{\phi(x)}$.

(i) As $|a|_2 < 1$, therefore $|x + a|_2 = |x|_2$.
(ii) Without loss of generality, assume line passes through the origin. But two other points (x_i, y_i); $i = 1, 2$ on the line $y = tx$ have colours blue and green respectively, say. But then $|y_1/x_1|_2 = |y_1/x_1|_2 = |t|_2$ is impossible.
(iii) Such segments on the boundary must be on the segment from $(0, 0)$ to $(1, 0)$ which are red and blue. But then it is clear.
(iv) This is clear because of the various possibilities RRB, RBB, RGG, RRG, BBG, BGG.

2.2 *Proof of Monsky's theorem*

The lemma easily implies Monsky's theorem as follows.

Counting the red-blue edges on the square, we are counting the interior edges twice and the boundary edges once. Thus, (iii) would be contradicted unless there is a complete triangle. But then a complete triangle has area A with $\phi(A) < 0$. So, if there are n triangles, then $\phi(A) = \phi(1/n) < 0$; that is, n is even.

2.3 *Proof of Stein's theorem*

Let us discuss Stein's generalization now. It can be seen that if a line segment made up of segments parallel to the axes has a blue end and a green end, then each of the individual segments has ends only coloured blue or green and, an odd number of them have both colours as ends. The key observation made already by Monsky is:

If a polyomino made up of standard squares as above is made up of n triangles of equal areas and, if an odd number of standard edges on its boundary have ends coloured blue and green, then $\phi(2A) \leq \phi(n)$, where A is the area of the polyomino.

The proof of this in turn depends on the following fact:

Let (x_i, y_i); $i = 0, 1, 2$ be the vertices of a triangle T where $(x_i, y_i) \in S_i$ with

$$S_0 = \{(x, y) : \phi(x), \phi(y) > 0\},$$

$$S_1 = \{(x, y) : \phi(x) \leq 0, \phi(y)\},$$

$$S_2 = \{(x, y) : \phi(y) < \phi(x), \phi(y) \leq 0\}.$$

Then, $\phi(\text{area})(T) \leq -\phi(2)$.

Proof. As translation by $(-x_1, -y_1)$ does not change areas, and $P_i - P_0 \in S_i$ for any $P_i \in S_i$, we may assume that $(x_0, y_0) = (0, 0)$.

Then, $\text{area}(T) = \frac{1}{2}\left|\det\begin{pmatrix} x_1 & y_1 \\ x_2 & y_2 \end{pmatrix}\right|$.

Now $\phi(x_1) \leq 0, \phi(y_1)$. Also $\phi(y_2) \leq 0$ and $\phi(y_2) < \phi(x_2)$.

Thus, $\phi(x_1y_2) < \phi(x_2y_1)$ and $\phi(x_1y_2) \leq 0$.

Hence $\phi(\text{area of } T) = \phi(1/2) + \phi(x_1y_2) \leq \phi(1/2) = -\phi(2)$.

To deduce Monsky's observation from this, look at a triangle of the dissection which has all three colours and let B denotes its area. Note that points in S_0, S_1, S_2 have different colours. Now, $nB = A$ and $\phi(B) \leq -\phi(2)$; that is, $\phi(A) - \phi(n) \leq -\phi(2)$. Therefore,

$$\phi(n) \geq \phi(2A).$$

Let us use Monsky's observation to prove Stein's result. We note that a standard (unit) edge with a blue end and a green end must be parallel to the X-axis and lies on a line whose height is odd. Therefore, on the border of each standard square, there is an edge with a blue end and a green end. Edges in the interior of the polyomino are adjacent to two standard squares whereas those on the boundary are adjacent to one standard square of the polyomino. As there is an odd number of standard squares, the above observation applies and, implies that $\phi(2A) \leq \phi(n)$. As $\phi(2A) \geq 1$, it follows that n must be a multiple of 2.

2.4 *Tiling rectangles by rectangles*

Let us discuss tiling integer rectangles with integer rectangles now. Can we tile a rectangle of size 28×17 by rectangles of size 4×7? At least the area of the smaller rectangle divides that of the larger one. But, in fact, we don't have a tiling for the following simple reason. Look at each row of the big rectangle. If we have managed to tile as required, then 17 would be a positive linear combination of 4 and 7. This is seen to be impossible. Thus, a necessary condition for tiling an $m \times n$ rectangle with $a \times b$ rectangles is that ab should divide mn and each of m, n should be expressible as positive linear combinations of a, b. Is this condition sufficient? Let us see.

Look at a 10×15 rectangle which we wish to tile with copies of a 1×6 rectangle. The two necessary conditions mentioned clearly hold true. However, this tiling is not possible! In fact, we claim that for an $a \times b$ rectangle to tile an $m \times n$ rectangle, it is also necessary that a must divide either m or n and b also must divide m or n. To see why, look at a possible tiling. Suppose $a > 1$ (if $a = b = 1$, there is nothing to prove). We colour the unit squares of the $m \times n$ rectangle with the different a-th roots of unity $1, \zeta, \zeta^2, \ldots, \zeta^{a-1}$ as follows. Think of the rectangle as an $m \times n$ matrix of unit squares and colour the (i, j)-th unit square by ζ^{i+j-2}. Since each tile (copy of the smaller rectangle used) contains all the a-th roots of unity exactly once, and as the sum $1 + \zeta + \cdots + \cdots \zeta^{a-1} = 0$, the sum of all the entries of the $m \times n$ rectangle must be 0. Therefore, $\sum_{i=1}^{m} \sum_{j=1}^{n} \zeta^{i+j} = 0$. But this sum is $(\sum_{i=1}^{m}, \zeta^i)(\sum_{j=1}^{n} \zeta^j) = 0$ which means one of these two sums must be 0. But $\sum_{i=1}^{m} \zeta^i = 0$ if, and only if, $\zeta^m - 1 = 0$; that is, $a|m$. The other sum is 0 if, and only if, $a|n$. Thus, this condition that a divides m or n is necessary and, by the same reasoning it is necessary for tiling that b divides m or n.

1	z	z^2	..	$1/z$	1	z	..	
z	z^2	..	$1/z$	1	z	..		
z^2	..							
..								
$1/z$								
1								
z								
..								

Looking at the above proof, it is also easy to see how to tile when these conditions hold good. That is, we have:

An $m \times n$ rectangle can be tiled with $a \times b$ rectangles if, and only if,

(i) *ab divides mn,*
(ii) *m and n are expressible as non-negative linear combinations of a and b,*
(iii) *a divides m or n and b divides m or n.*

It is an interesting exercise to generalize this to any dimension!

2.5 *Tiling squares and orthogonal polygons by similar copies of a rectangle*

Let us discuss a slightly different tiling problem which has an extremely surprising answer. Suppose $t < 1$ is a fixed positive real number and we try to tile the unit square by rectangles which are similar to a $1 \times t$ rectangle. This means we are allowed to use $a \times at$ rectangles for any a. We look at an example first. Suppose we have put on the upper part of the unit square a $1 \times t$ rectangle (that is, the base of the rectangle is at a height $1 - t$ from the base of the square). If we are able to tile the bottom half of the square by placing d identical rectangles from the left to the right which are all similar to the $1 \times t$ rectangle. Each of these rectangles then have width $t(1 - t)$ since their height is $1 - t$. Therefore, the total width is $1 = dt(1 - t)$.

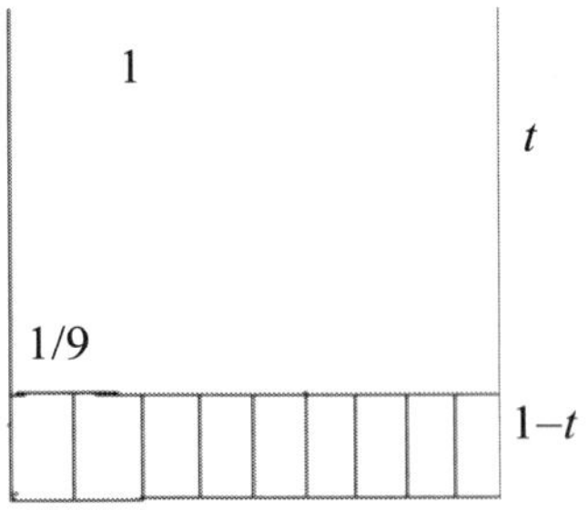

d=9, t/1=(1/9)/(1−t)

For each $d \geq 4$, there are two real values of $t = \frac{1 \pm \sqrt{1-4/d}}{2} < 1$ satisfying $dt^2 - dt + 1 = 0$; each gives a tiling of the unit square by d rectangles of size $(1-t) \times t(1-t)$ and one rectangle of size $1 \times t$. This situation might look very special but in reality it turns out to be typical! We have:

A square can be tiled with rectangles similar to $1 \times t$ if and only if t is algebraic (a root of a non-zero integral polynomial), and all its conjugates (the roots of its minimal polynomial) have positive real part.

For instance, this implies bizarre things like:

A square cannot be tiled by rectangles similar to a $1 \times \sqrt[3]{2}$ rectangle. A square can be tiled by rectangles similar to a $1 \times (1 + \sqrt[3]{2})$ rectangle.

The above result holds more generally. Call a polygon to be *orthogonal* if its sides are parallel to the axes. We have:

For $t > 0$, an orthogonal polygon can be tiled by rectangles similar to a $1 \times t$ rectangle if, and only if, t is algebraic and all its conjugates have positive real part.

The proof interestingly proceeds via continued fractions.

2.6 *Rephrasing in terms of continued fractions*

The condition of algebraicity of t with all conjugates of t having positive real part turns out to be equivalent to the condition that 1 has a continued fraction of the type

$$1 = c_0 t + \cfrac{1}{c_1 t + \cfrac{1}{c_2 t + \cdots \frac{1}{c_n t}}}$$

for some positive rational numbers $c_0, \ldots, c_n$.

2.7 *Illustration*

Let us illustrate how this latter condition provides a tiling.

Consider $t = 2 - 5^{1/3}$. Its minimal polynomial is $T^3 - 6T^2 + 12T - 3$.

The 'even' part is $E(T) = 6T^2 + 3$ and the 'odd' part is $O(T) = T^3 + 12T$. Note that $E(t) = O(t)$. The rational function $E(T)/Q(T)$ has the continued fraction expansion

$$\frac{6T^2 + 3}{T^3 + 12T} = \cfrac{1}{\frac{1}{6}T + \cfrac{1}{\frac{12}{23}T + \cfrac{1}{\frac{23}{6}T}}}.$$

Write $c_0 = 0, c_1 = \frac{1}{6}, c_2 = \frac{12}{23}, c_3 = \frac{23}{6}$. Consider the numbers $a = c_3 t, b = c_2 t a$, $c = c_1 t + c_1 c_2 t^2 a = c_1 t(1 + b)$. With the help of these numbers, consider rectangles of sizes $1 \times a$, $a \times b$ and $(1 + b) \times c$.

Note that the sum $a + ab + (1 + b)c = (a + c)(1 + b)$ of their areas works out to $(4t + t^3/3)(1 + 2t^2) = (1 + 2t^2)^2$ (as $t^3 - 6t^2 + 12t - 3 = 0$). Thus, we may fit the 3 rectangles into a square with side $1 + 2t^2$ by putting the $1 \times a$ rectangle at the left lower corner with 1 as height, the $a \times b$ rectangle right on top if it (sharing the side of length a with the lower rectangle), the rectangle of height $1 + b$ and width c on the right of these two. Observe that each of the rectangles has eccentricity which is a rational multiple of t. Thus, they may all be tiled by rectangles similar to a $1 \times t$ rectangle!

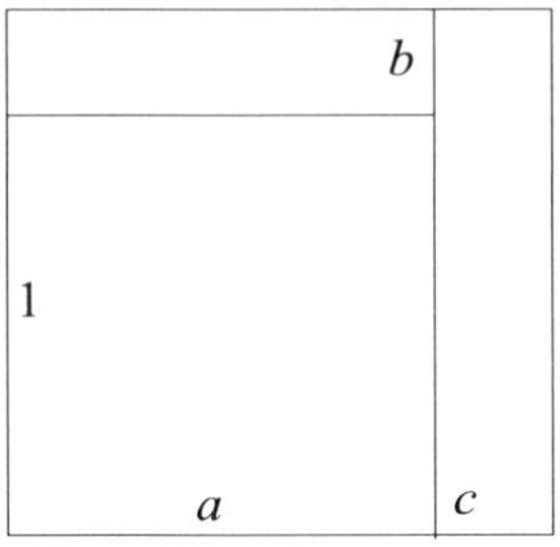

A slightly different tiling problem which can be solved using simple linear algebra has an interesting answer.

An orthogonal polygon can be tiled by similar triangles if, and only if, the triangles are right triangles or the angles are rational multiples of π.

3. Conway's tiling group

In order to study several tiling problems in a united way, John Conway introduced the so-called tiling group. This group not only provides necessary conditions for finite tiling problems but gives an algorithm to find a tiling if there is one. Let us define this now without further ado.

Consider the circle group S^1 of complex numbers of absolute value 1 and, the free product $G = S^1 * S^1$. This is just the group of all expressions $h_1 v_1 h_2 v_2 \ldots h_r v_r$ where h_i's come from the first factor and v_i's come from the second. The product operation in this group is simply concatenation and the group operations on the h's and the v's individually and cancelling off to get a reduced expression. The point this group is relevant is as follows.

Look at a simple rectilinear polygon R with edges parallel to the axes and look at the closed path running along its boundary in the anti-clockwise direction. Let the origin be on the boundary of the polygon and think of the path as starting at the origin. Now, we can naturally associate an element of the group G as follows. For each horizontal distance d covered in the positive X-direction, consider $h(d)$ to be the corresponding element $e^{2id\pi}$ of the first S^1. For any vertical distance l in the positive Y-direction, associate the element $v(l)$ of the second S^1 which is $e^{2il\pi}$. Thus, the anti-clockwise path along the boundary of R starting from the origin and ending there looks like an element $h(s_1)v(t_1)h(s_2)v(t_2)\cdots v(t_n)$ where $s_1 + \cdots + s_n = 0 = t_1 + \cdots + t_n$.

For example, along a rectangle of size $\alpha \times \beta$ with left bottom corner at the origin, the corresponding element of G is $h(\alpha)v(\beta)h(-\alpha)v(-\beta)$; one often writes this 'commutator element as $[h(\alpha), v(\beta)]$. In general, call the element of G corresponding to R as $\alpha(R)$.

One has the following simple, beautiful result giving a necessary criterion for tiling:

Lemma. *If R can be tiled by rectangles each of which has at least one integer side, then $\alpha(R) = e$ in G. In particular, if R is an $l \times b$ rectangle tileable as above, then either $l \in \mathbf{Z}$ or $b \in \mathbf{Z}$.*

Proof. A tiling of R will present the element $\alpha(R)$ as a product of 'lassos'; that is, expressions of the form $x[h(s), v(t)]x^{-1}$. Here, the commutators $[h(s), v(t)]$ come from the tiles. This can be proved by induction on the number of tiles. The inductive step is done by taking a tile touching the boundary of R as in the figure. Now, each lasso is trivial because one of s, t is integral (!) Therefore, $\alpha(R)$ itself is trivial.

The last assertion follows because if $e = \alpha(R) = [h(l), v(b)]$, then either $h(l) = e$ or $v(b) = e$; that is, $l \in \mathbf{Z}$ or $b \in \mathbf{Z}$.

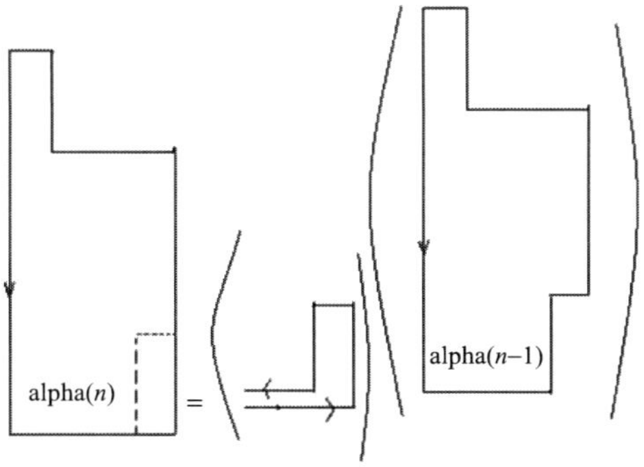

Remark. The lemma gives only a necessary condition for tileability; however it is not sufficient as seen by looking at the boundary word

$$h(2/3)v(1/3)h(-1/3)v(1)h(-2/3)v(-1/3)h(1/3)v(-1)$$

which is trivial. The corresponding polygon cannot be tiled.

3.1 *Tile group and tile homotopy group*

Now, we define the so-called tile group determined by a given set of tiles. Let us call a *cell* the interior and boundary of a unit square whose vertices are lattice points. A *region* is a finite, connected set of closed cells. The topological boundary of a cell consists of its four edges, oriented anti-clockwise. The boundary ∂R of a region R is formed by taking the set of all edges in the boundary of all cells in it, and discarding any edge that occurs twice with opposite orientations. A region is simply-connected in the usual topological sense if, and only if, its complement in the plane is connected and if its boundary edges can be ordered to form a simple closed path (so, one may take the latter as the definition of a simply-connected region).

For a simply-connected region R, a simple closed path bounding it is specified by its leading (first) edge. Thus, one writes $\partial R(e)$ for a oriented boundary of R with leading edge e. Its first vertex is called its base point. One writes an oriented boundary

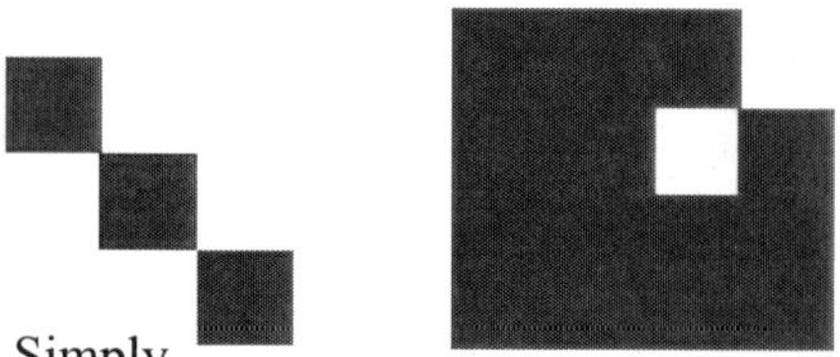

Simply connected

Not simply connected

as a word in the h's and the v's as before but reading from right to left. To be clear, we write $R = h(1)$, one horizontal move to the right and write $U = v(1)$, a vertical move upwards. We consider simply-connected regions made up of cells to be our tiles. If there are n cells, we call it an n-tile. Thus, tiling a region R with a set Σ of tiles means covering each cell of R exactly once using a set of tiles obtained by translating tiles in Σ. Consider the free group F in the two elements R, U.

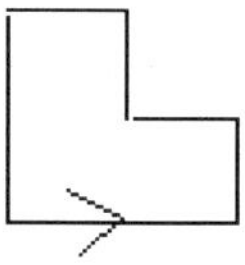

dR(e_1)=U^{-2}R^{-1}UR^{-1}UR^2

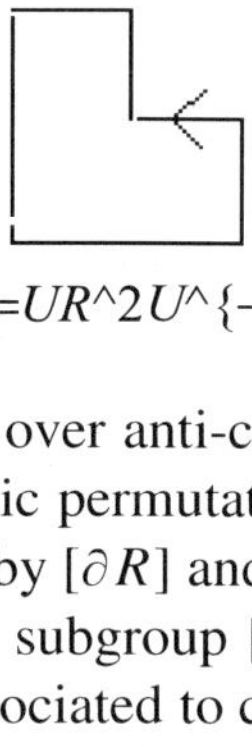

dR(e_2)=UR^2U^{-2}R^{-1}UR^{-1}

Note that $\partial R(e)$, as e varies over anti-clockwise oriented edges of ∂R, are conjugate in F as the words are cyclic permutations of each other. One denotes the corresponding conjugacy class in F by $[\partial R]$ and calls it the combinatorial boundary. It can be proved that the commutator subgroup $[F, F]$ coincides with the 'cycle group' C consisting of all words in F associated to closed directed paths in the square lattice.

The *tile group* $T(\Sigma)$ associated to a set of tiles $\Sigma = \{R_i\}_{1 \leq i \leq m}$ is the normal subgroup generated by the corresponding conjugacy classes $[\partial R_i]$; $1 \leq i \leq m$. Therefore,

$$T(\Sigma) = \langle g \partial R_i(e_i) g^{-1} : g \in F, 1 \leq i \leq m \rangle.$$

Notice that $T(\Sigma) \leq C$, the cycle group. In analogy with the edge path group in homotopy theory, the quotient group $h(\Sigma) = C/T(\Sigma)$ is called the *tile homotopy group*.

The physical meaning of this tiling group is that in the set of all allowable closed paths, the tile homotopy group represents paths which can be deformed to the empty path by picking up or laying down tiles.

The following statement seems intuitive but requires some care to give a rigorous proof using induction on the number of tiles.

A necessary condition for a simply-connected region R to have a tiling by translates of a finite set Σ of tiles is that the combinatorial boundary $[\partial R]$ be contained in the tile group $T(\Sigma)$. In other words, considered in the tile homotopy group corresponding to the tile set Σ, the combinatorial boundary of R is trivial.

3.2 *Tiling by arrays of hexagonal arrays*

There are other tiling problems which can be converted into tiling problems in the square lattice as above. For instance, when we consider the hexagonal tiling of the plane, we could look at the triangular array T_N of $1 + 2 + \cdots + N$ hexagons which we wish to tile by congruent copies of T_2 or by three-in-line hexagons L_3. Using the tiling group, the interesting fact which emerges is:

T_N can be tiled with congruent copies of T_2 if, and only if, $N \equiv 0, 2, 9, 11$ mod 12.
T_N cannot be tiled with congruent copies of L_3.

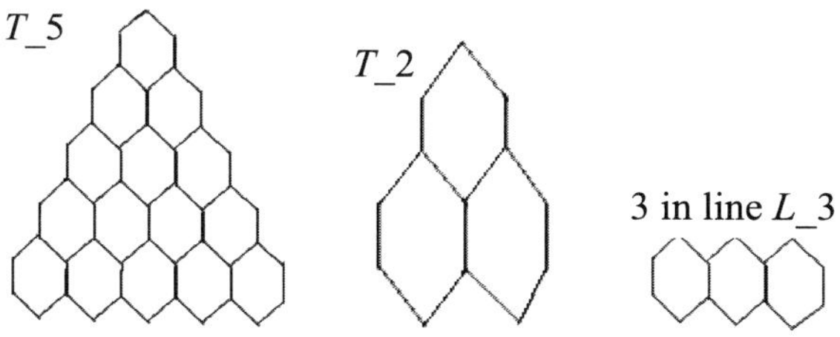

One may view the region to be tiled as a *staircase* of squares and the tiles T_1 and L_3 as squares as well.

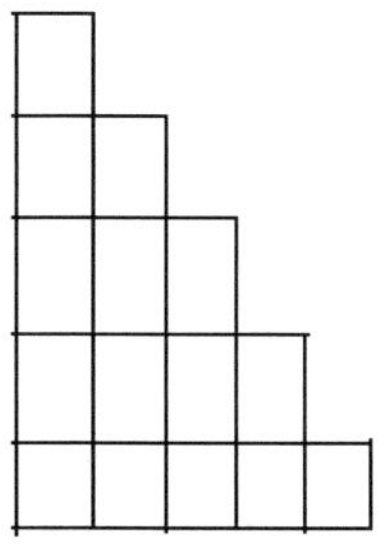

Staircase region T_5

A word representing the boundary of the staircase form of T_N starting from the right bottom right corner is $\partial T_N = R^N U^{-N}(R^{-1}U)^N$ which is read from right to left as before. Then, we also have the two congruent copies of T_1 to have boundaries $\partial R_1 = R^2U^{-2}(R^{-1}U)^2$, $\partial R_2 = (RU^{-1})^2R^{-2}U^2$. The three congruent forms of L_3 are R_3, R_4, R_5 which have $\partial R_3 = [R^3, U^{-1}]$, $\partial R_4 = [R, U^{-3}]$, $\partial R_5 = (RU^{-1})^3(R^{-1}U)$. The tile groups of all the R_i's above turn out to be contained in a certain special normal subgroup H of $F = \langle R, U\rangle$. This subgroup H is determined by the property that the (undirected) Cayley diagram of F/H is the infinite planar graph that tiles the plane with hexagons and triangles (see figure below). Recall that, in general, a Cayley diagram is a graph which depicts a group; there is one node for each element, and arrows show how the generators act on the elements of the group. The subgroup H can be explicitly defined as the normal subgroup generated by the elements R^3, U^3 and $(U^{-1}R)^3$. As a matter of fact, the quotient group F/H is the plane crystallographic group $p3$ which is one of the 17 such groups. The fact that H contains the tile groups of $T(R_1), T(R_3)$ follows simply by checking that the representative generators ∂R_i's give closed paths at I on the Cayley diagram $\mathcal{G}$ of F/H.

In fact, H also contains all $\partial T_N = R^N U^{-N}(R^{-1}U)^N$ for $N \equiv 0, 2 \bmod 3$; this is because H contains R^3, U^3 and $(R^{-1}U)^3$ and so, ∂T_N would be in H provided ∂T_i is in H for $N \equiv 0, 2 \bmod 3$. As it is easily seen that $\partial T_2, \partial T_3$ are in H, our assertion follows.

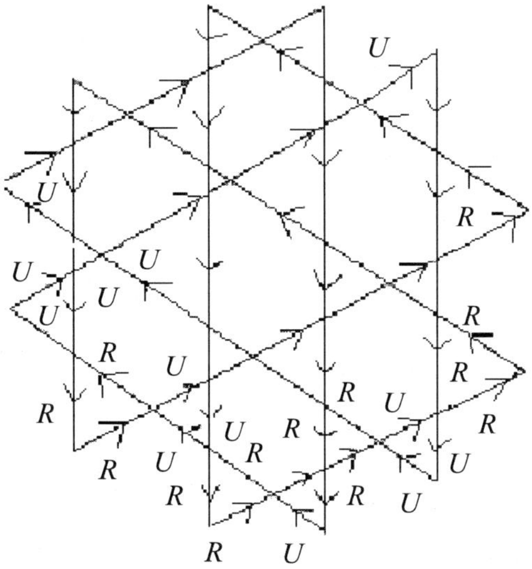

Cayley diagram *G*(*F*/*H*)

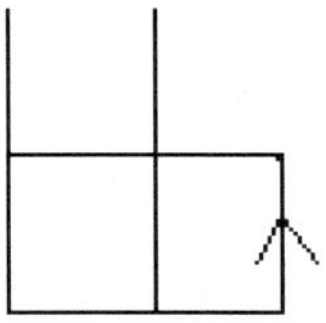
dR 1=*R*^2*U*^{−2}(*R*^{−1}*U*)^2

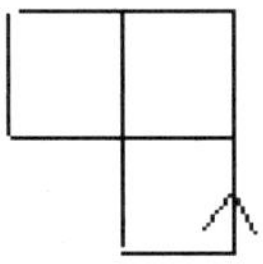
*dR*_2=(*RU*^{−1})^2*R*^{−2}*U*^2

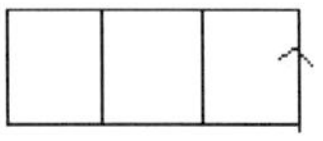
*dR*_3=*R*^{3}*U*^{−1}*R*^{−3}*U*

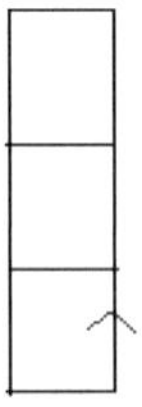
*dR*_4–*RU*^{−3}*R*^{−1}*U*^3

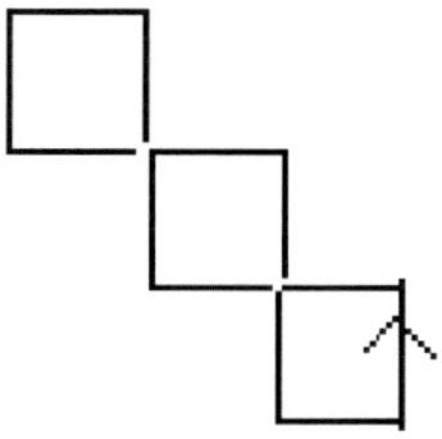

$dR\_5=(RU^\wedge\{-1\})^\wedge3(R^\wedge\{-1\}U)^\wedge3$

The key concept needed to answer the above two tiling problems is that of the *winding number* $w(P, s)$ of a closed, oriented path P in $\mathcal{G}$ around a (hexagonal or triangular) cell s in a tiling; it is given by $\frac{1}{2\pi i}\int_P \frac{1}{z-x_s}dz$ where x_s is any point inside s. This is an additive function when we consider paths starting at the same point. The winding number thus gives a homomorphism: $H \to \mathbf{Z}$ which assigns to every word in H, the value $w(P, s)$ for the corresponding closed oriented path P. Let us now prove the above assertions:

Theorem 3.1.

(i) T_N *can be tiled with congruent copies of* T_2 *if, and only if,* $N \equiv 0, 2, 9, 11$ *mod* 12.

(ii) T_N *cannot be tiled with congruent copies of* L_3.

Proof. (i) Firstly, if T_2 covers 3 cells, it can cover T_N only if 3 divides $1+2+\cdots+N$ which gives $N \equiv 0, 2 \bmod 3$. We assume that T_N has been tiled by congruent copies of T_2 and show now that ∂T_N is not in the tile groups $T(R_1)$, $T(R_2)$ when $N \equiv 3, 5, 6, 8$ mod 12, which gives a contradiction.

Consider the homomorphism $\phi : H \to \mathbf{Z}$ given by $W \mapsto \sum_s w(W, s)$ where the sum varies over the finite cells enclosed by a word W.

Using the expressions

$$\partial R_1 = R^2U^{-2}(R^{-1}U)^2,$$

$$\partial R_2 = (RU^{-1})^2R^{-2}U^2$$

it is easy to compute and find $\phi(\partial R_1) = 1$ and $\phi(\partial R_2) = -1$.

Now, $\partial T_N = \prod_{i=1}^{m} W_i(\partial R_t)^{\epsilon_i}W_i^{-1}$ where $t = 1$ or 2.

Thus, $\phi(\partial T_N) \equiv m \bmod 2$.

As H contains $\partial T_N = R^NU^{-N}(R^{-1}U)^N$ for the N above, we also have $\phi(\partial T_N) = [(N+1)/3]$.

Thus, we have

$$m \equiv [(N+1)/3] \bmod 2.$$

On the other hand, another homomorphism ψ on H is defined by viewing each word in H as a closed oriented path in the square lattice, starting at the origin and assigning the sum of the winding numbers about cells in the square lattice.

Thus, $\psi(\partial R_i) = 3$ for $i = 1, 2$.

Also, $\psi(\partial T_N) = 1+2+\cdots+N = \binom{N+1}{2}$.

Hence $\binom{N+1}{2} \equiv 3m \equiv m \bmod 2$ as $\partial T_N = \prod_{i=1}^{m} W_i(\partial R_t)^{\epsilon_i} W_i^{-1}$ where $t = 1$ or 2. Hence

$$m \equiv \binom{N+1}{2} \bmod 2.$$

Therefore, $N \not\equiv 3, 5, 6, 8 \bmod 12$.

To complete (i), we need to construct tilings of T_N by T_2 whenever $N \equiv 0, 2, 9, 11 \bmod 12$.

It is easy to construct tilings of T_2, T_9, T_{11} and T_{12} by T_2 and not difficult to extend a tiling of T_N to one of T_{12+N}.

Proof. (ii) We just indicate how this follows similarly to the above. Once again, we assume a tiling of T_N exists by tiles congruent to L_3 and easily deduce that $N \equiv 0, 2 \bmod 3$ as earlier. Once again, we consider the homomorphism θ on H counting the sum of all winding numbers of a path around triangles marked U in $\mathcal{G}$. Then, it is easily seen that $\phi(\partial R_i) = 0$ for $i = 3, 4, 5$ whereas $\theta(\partial T_N) = [(N+1)/3]$. Thus, if $\partial T_N = \prod_{i=1}^{m} W_i(\partial R_t)^{\epsilon_i} W_i^{-1}$ where $t = 3, 4$ or 5, then $[(N+1)/3] = 0$, which is impossible. Thus, we obtain a contradiction to the assumption that T_N can be tiled with the tiles congruent to L_3.

4. Hajós's proof–Rédei's theorem

Finally, we come to the proof of the Minkowski conjecture. As we saw in the beginning, Hajós reformulated it in terms of groups and proved it. His proof used zero divisors in integral group rings. Rédei tried to remove the reference to group rings and succeeded after 25 years of work (!) to give a proof of the following more general result:

Theorem (Rédei). *Suppose $G = A_1 A_2 \cdots A_n$ is a factorization of a finite, abelian group where $A_1, \ldots, A_n$ are subsets each of which contains the identity and has a prime number of elements. Then, at least one A_i is a subgroup.*

Hajós's theorem asserting that "*In any factorization of a finite abelian group by cyclic subsets, at least one of the factors is a subgroup*", turns out to be equivalent to its special case when the factors have prime cardinalities; this is actually easy to show. Therefore, Rédei's theorem indeed implies Hajós's theorem.

We indicate a proof of Rédei's theorem which makes some mild use of group rings though. The procedure is to first prove Rédei's theorem when the factors are cyclic and G is a p-group (so, this is itself a special case of Hajós's theorem. Following that, one proves Rédei's theorem for cyclic p-groups (this is also easy) and then the noncyclic group of order p^2 is treated (here the proof is involved). By an inductive argument, the theorem is deduced in general.

We start with two easy observations:

Observation. *If $G = AB$ is a factorization of a finite abelian group G where $A = \{e, a, a^2, \ldots, a^{p-1}\}$ for a prime p and B contains identity, then $G = A'B$ is also a factorization where $A' = \{e, a^k, a^{2k}, \ldots, a^{(p-1)k}\}$ for any k coprime to p.*

Observation. *Let H be a proper subgroup of an abelian group G. Suppose $H = C_1 \cdots C_r$ and $G/H = D_1^* \cdots D_s^*$ are factorizations. Then, $G = C_1 \cdots C_r D_1 \cdots D_s$ is a factorization where D_i is any set of H-coset representatives of D_i^*.*

From these observations, we have:

Lemma. *Let G be an abelian p-group and A a subset for which $|\langle A \rangle| = p^{|A|}$ and $|\langle B \rangle| \geq p^{|B|}$ for all $B \subset A$. Then, for each $a \in A$, there exists a natural number $s(a)$ which is a power of p, such that*

$$\langle A \rangle = \prod_{a \in A} \{e, a^{s(a)}, \ldots, a^{(p-1)s(a)}\}$$

is a factorization in which at least one factor is a subgroup.

Proof. It holds when $|A| = 1$ and we apply induction on $n = |A|$. The number $h(A) := \prod_{a \in A} O(a)$ satisfies $h(A) \geq p^n$ with equality when each element of A has order p. Thus, for a given n, we may apply induction on $h(A)$. If it happens that $|\langle B \rangle| > p^{|B|}$ for all $B \subset A, B \neq A$, we may replace one of the elements of A by its p-th power and obtain a subset A' which satisfies $\langle A' \rangle = \langle A \rangle$ and $h(A') < h(A)$. Thus, by induction hypothesis, the result holds for A' and hence for A. Therefore, we assume that there is a non-empty proper subset B of A for which $|\langle B \rangle| = p^{|B|}$. Once again, induction hypothesis implies that $\langle B \rangle$ has a factorization $\prod_{b \in B} \{e, b^{s(b)}, \ldots, b^{(p-1)s(b)}\}$ where $s(b)$ is some power of p. But then $\bar{G} := \langle A \rangle / \langle B \rangle$ has subsets $\bar{A} := \{aB : a \in A \backslash B\}$ which satisfy $\langle \bar{A} \rangle = p^{|\bar{A}|}$ and $\langle \bar{B} \rangle \geq p^{|\bar{B}|}$ for all subsets $\bar{B}$ of $\bar{A}$. By induction on $|A|$, we have a factorization of $\langle \bar{A} \rangle$ of the type in the lemma and, using the factorizations of $\langle B \rangle$ and of $\bar{G} = \langle A \rangle / \langle B \rangle$, we deduce a factorization of the desired form for G by the second observation.

The first observation above (which is a sort of 'replacement principle') and the lemma, we can deduce:

Hajós's theorem for p-groups. *Let $G = A_1 A_2 \cdots A_n$ be a factorization of a finite abelian p-group, where $A_i = \{e, a_i, \ldots, a_i^{p-1}\}$. Then, at least one A_i is a subgroup.*

Proof. Applying the lemma to $A = \{a_1, \ldots, a_n\}$, we get, for each $i \leq n$, a power $s(i)$ of p and a subset $A_i' = \{e, a_i^{s(i)}, \ldots, a_i^{(p-1)s(i)}\}$ such that $G = A_1' A_2' \cdots A_n'$ is a factorization where at least one A_i' is a subgroup. Suppose $s(i) \neq 1$ for some i. We rename the indices so that $s(i) \neq 1$ for all $i \leq m$ and $s(j) = 1$ for all $m < j \leq n$. Then $G = A_1' A_2' \cdots A_m' A_{m+1} \cdots A_n$ is a factorization. Thus, the element

$$a_1 a_2 \cdots a_m = a_1^{s(1)t(1)} \cdots a_m^{s(m)t(m)} a_{m+1}^{t(m+1)} \cdots a_n^{t(n)}$$

with $0 \leq t(i) \leq p - 1$. Therefore,

$$e = a_1^{s(1)t(1)-1} \cdots a_m^{s(m)t(m)-1} a_{m+1}^{t(m+1)} \cdots a_n^{t(n)} \cdots \spadesuit$$

As $(s(i)t(i), p) = 1$ for all $i \leq m$, the first observation allows us to replace the factor A_i by $A_i^* := \{e, a_i^{s(i)t(i)-1}, \ldots a_i^{(p-1)s(i)t(i)-1}\}$ for each $i \leq m$. The equality $\spadesuit$ violates such a factorization. Thus, we have $m = 0$; that is, $s(i) = 1$ for all $i \leq n$ and we are done. The proof is complete.

4.1 *Group rings for factorization*

We now introduce characters of abelian groups and group rings in order to discuss Rédei's theorem for cyclic p-groups. A *character* of an abelian group is simply a homomorphism from the group to the unit circle group S^1. If χ_1, χ_2 are two characters of an abelian group G, then their product $\chi_1\chi_2$ can be defined to be the character which sends any element g to the product $\chi_1(g)\chi_2(g)$. Thus, the set $X(G)$ of characters itself forms an abelian group with the identity element being the constant map 1. For instance, the characters of S^1 itself can all be found; they are the n-th power maps for any integer n. Thus, $X(S^1) \cong \mathbf{Z}$. Similarly, the character group $X(\mathbf{Z}) \cong S^1$ because the characters of $\mathbf{Z}$ are the maps $n \mapsto e^{in\theta}$ for any element $e^{i\theta}$ of S^1. The character group $X(\mathbf{Z}_n)$ of the cyclic group $\mathbf{Z}_n$ of order n is again isomorphic to itself because the characters can be seen to be the maps $\bar{r} \mapsto e^{2ik\pi/r}$ for $k = 0, 1, \ldots, n-1$.

The main properties of characters which play crucial roles are:

(i) If $G = A_1A_2\cdots A_n$ is a factorization, then
$$\chi(G) = \chi(A_1)\chi(A_2)\cdots\chi(A_n)$$
for all $\chi \in X(G)$ where $\chi(G) := \prod_{g\in G}\chi(g)$ etc.;

(ii) If $A, B \subset G$, where G is a finite, abelian group then $A = B$ if and only if, $\chi(A) = \chi(B)$ for all $\chi \in X(G)$.

If G is a finite, abelian group, the integral group ring $\mathbf{Z}[G]$ is defined to be the set of all formal integral linear combinations $\sum_{g\in G} a_g g$ (this can also be thought of as a function from G to $\mathbf{Z}$ in an obvious manner). One may add elements of $\mathbf{Z}[G]$ by adding the corresponding coefficients (same thing as addition of functions). One may multiply in $\mathbf{Z}[G]$ as:

$$\left(\sum_{g\in G} a_g g\right)\left(\sum_{g\in G} b_g g\right) := \sum_{g\in G}\left(\sum_{xy=g} a_x b_y\right) g.$$

This corresponds to convolution of functions. Thus, $\mathbf{Z}[G]$ is a commutative ring with unity, called the integral group ring of G. The characters of G extend to homomorphisms from $\mathbf{Z}[G]$ to S^1 in an obvious manner :

$$\chi\left(\sum_{g\in G} a_g g\right) = \prod_{g\in G}\chi(g)^{a_g}.$$

The main advantage of considering the group ring in our context is that $G = AB$ is a factorization of a finite, abelian group if, and only if, $G = AB$ in $\mathbf{Z}[G]$. Equivalently, $\chi(G) = \chi(A)\chi(B)$ for all $\chi \in X(G)$. A further equivalent condition is useful and can be proved easily:

$G = AB$ is a factorization of a finite, abelian group if, and only if, $|G| = |A||B|$ and for every nontrivial character χ, one of $\chi(A)$ and $\chi(B)$ is zero.

From the above, one has the following generalization of observation 1:

4.2 *Proof of Rédei's theorem*

General replacement principle. *Let A, A' be subsets of a finite, abelian group G of equal cardinality such that whenever $\chi(A) = 0$, we also have $\chi(A') = 0$. Then, in all factorizations $G = AB$, one can replace A by A' and get a factorization $G = A'B$.*

Note that the group ring is not explicitly appearing in the statement above. The above general principle implies also the special replacement principles:

(I) *Let $A = \langle a \rangle$ be a subset of cardinality p in a finite, abelian p-group G. Let $e \in A$ and $(p, k) = 1$. Then, in each factorization of G, A can be replaced by $A' = \{e, a^k, \ldots, a^{(p-1)k}\}$.*

(II) *Let $|A| = p$, a prime. Assume also that each nontrivial element of A has order p. Then, in any factorization involving A, the set A can be replaced by a cyclic subset; that is, a subset of the form $\{e, a, a^2, \ldots, a^r\}$.*

These can be used to prove of Rédei's theorem for abelian p-groups.

To prove Rédei's theorem for non-p-groups, the following fact about sum of roots of unity will be used; it is a nice exercise [24].

Let ζ be a primitive p-th root of unity and n be a multiple of the prime p. If $1 + \zeta^{a_1} + \cdots + \zeta^{a_s} = 0$ for some $0 \le s \le p-1$, then $s = p$ and there is a primitive p-th root of unity ρ such that $\{\zeta^{a_1}, \ldots, \zeta^{a_{p-1}}\} = \{\rho, \rho^2, \ldots, \rho^{p-1}\}$.

The proof of Rédei's theorem for non-p-groups proceeds by induction on the order and another replacement results deduced from the general replacement result and facts on group rings and characters. We do not give the details but, in order to give a flavour of the proof and to indicate how group rings and characters come into the proof, we prove one key result now. Before stating it, we note that for a general finite, abelian group G and a prime p dividing $O(G)$, G is $G(p)\bar{G}(p)$ where $G(p)$ is the p-component of G and $\bar{G}(p)$ is the complementary direct factor. Therefore, each element $g \in G$ has a unique expression $g = g(p)\bar{g}(p)$. The result is:

Let A be a factor of G such that $|A| = p$ is the least prime factor of $O(G)$ and assume $e \in A$. Then, in any factorization of G where A occurs, A can be replaced by $A' := \{a(p)(\bar{a}(p))^{s(a)} : a \in A\}$ for arbitrary choices of $s(a)$.

Proof. Write $G = AB$, a factorization such that $e \in B$. Suppose there is no character of G which vanishes on A. Then, $\chi(B) = 0$ for each non-principal character χ of G. Therefore, each character of G vanishes on the element $|B|G - |G|B$ of the group ring (note that this is true for the principal character also). This means the elements $|B|G = |G|B$ in $\mathbf{Z}[G]$. This is a contradiction as the coefficients o e are different. Therefore, there is a character χ such that $\chi(A) = \sum_{a \in A} \chi(a) = 0$. Let n be the smallest positive integer so that each $\chi(a)$ is a primitive n-th root of unity. As $n | O(G)$, the least prime divisor of n is $\ge p$. By the above fact on vanishing sums of roots of unity, there is a primitive p-th root of unity, say ζ, such that

$$\{\chi(a) : a \in A\} = \{\zeta^i : 0 \le i < p\}.$$

Thus, $|A'| = p$ and $\chi(A) = 0$ implies $\chi(A') = 0$. By the general replacement principle mentioned earlier, A can be replaced by A' in any factorization.

5. Reference to music

An extremely remarkable fact is that some of the tiling theorems have been discovered independently in the context of music! For instance, Dan Tudor Vuza was a musician who proved/rediscovered many results over 10 years until the mathematical community became aware of his contributions. Let us just give a very brief glimpse of how the connection with music arises.

If the cyclic group $\mathbf{Z}_n$ is tiled by translates of some subset A; say,

$$\mathbf{Z}_n = A \sqcup (A + b_1) \sqcup (A + b_2) \sqcup \cdots = A \oplus B$$

with $B = \{0, b_1, b_2, \ldots\}$, then one says that we have a rhythmic canon. Musically, this can be rendered as a canon (say, with percussion instruments) repeating with period n, playing any motif (modelled as a collection of integers corresponding to the beats played) that reduces to A modulo n, beginning each instance of this motif on beats congruent mod n to the b_i's. One calls A as the inner rhythm and B as the outer rhythm. Vuza defined a special type of tiling (which corresponds to what is now known as a Vuza canon) wherein $\mathbf{Z}_n = A \oplus B$ with $A + r \neq A$, $B + r \neq B$ for all $1 < r < n$. Such tilings are much rarer. For instance, the smallest n for which they occur is $n = 72$. Referring to those cyclic groups $\mathbf{Z}_n$ for which they do not occur as Hajós groups, Vuza proved the following classification theorem and Vuza even gave an algorithm to determine a Vuza canon for any non-Hajós group:

The Hajós groups are those cyclic groups whose order is of one of the forms p^k, $p^k q$, $p^2 q^2$, $p^2 qr$, $pqrs$ where p, q, r, s are distinct primes.

5.1 *A conjecture*

In the context of music or of harmonic analysis, the fundamental open conjecture is Fuglede's conjecture. It is not always true as proved by Terence Tao in 2003 when he found a certain counter-example in a high dimension. To state it, let us define a couple of things. Given a tiling $\mathbf{Z}_n = A \oplus B$, one can lift it to a tiling of $\mathbf{Z}$ as $\mathbf{Z} = A \oplus (B + n\mathbf{Z})$. In turn, this gives a tiling of $\mathbf{R}$ itself as $\mathbf{R} = (A + [0, 1)) \oplus (B + n\mathbf{Z})$. The conjecture asserts:

A finite subset A gives a tiling by translations as above if, and only if, it is spectral; that is, there is some $\Lambda \subset [0, 1)$ with $|A| = |\Lambda|$ and $e^{2i\pi(\lambda_j - \lambda_k)}$ is a root of the polynomial $\sum_{r \in A} X^r$ for all $\lambda \neq \lambda_k \in \Lambda$.

Acknowledgements

It is a pleasure to acknowledge the comments and suggestions of the reviewer. I also thank Vidur (12) for producing the figures in a jiffy.

References

[1] E. Amiot, New perspectives on rhythmic canons and the spectral conjecture, *Journal of Mathematics and Music*, **3(2)** (July 2009) 71–84.

[2] E. Amiot, Rhythmic canons and Galois theory, Colloquium on Mathematical Music theory, edited by H. Fripertinger and L. Reich, *Grazer Math. Ber.*, ISSN 10167692 Bericht Nr., **347** (2005) 1–25.

[3] J. H. Conway and J. C. Lagarias, Tiling with Polyominoes and Combinatorial Group Theory, *Journal of Combinatorial Theory A*, **53** (1990) 183–208.

[4] E. M. Coven and A. Meyerowitz, Tiling the Integers with Translates of One Finite Set, *Journal of Algebra*, **212** (1999) 161–174.

[5] C. Freiling, R. Hunter, C. Turner and R. Wheeler, Tiling with squares and anti-squares, *The American Mathematical Monthly*, **107(3)** (Mar., 2000) 195–204.

[6] C. Freiling, M. Laczkovich and D. Rinne, Rectangling a Rectangle, *Discrete Comput Geom.*, **17** (1997) 217–225.

[7] G. Hajós, Uber einfache und behrfache Bedeckung des n-dimensionalen raumes mit einenwurfelgitter, *Math. Zeit*, **47** (1942) 427–467.

[8] F. Jedrzejewski, Tiling the integers with aperiodic tiles, *Journal of Mathematics and Music*, **3(2)** (July 2009) 99–115.

[9] O. H. Keller, Uber die luckenlose einfullung des raumes mit wurfeln, *J. Reine. Angew. Math.*, **163** (1930) 231–248.

[10] R. Kenyon, A note on tiling with integer-sided rectangles, *Journal of Combinatorial Theory A*, **74** (1996) 321–332.

[11] C. Kenyon and R. Kenyon, Tiling a polygon with rectangles, *Proceedings of the 33rd IEEE conference on Foundations of Computer Science (FOCS)*, (1992) 610–619.

[12] M. Kolountzakis, Lattice tilings by cubes: whole, notched and extended, *Integers, the electronic journal of combinatorics*, **5** (1998) #R14.

[13] M. N. Kolountzakis and M. Matolcsi, Algorithms for translational tiling, *Journal of Mathematics and Music*, **3(2)** (July 2009) 85–97.

[14] M. Laczkovich, Tilings of triangles, *Discrete Mathematics*, **140** (1995) 79–94.

[15] J. C. Lagarias and P. W. Shor, Keller's cube-tiling conjecture is false in high dimensions, *Bull. Amer. Math. Soc., N. S.*, **27** (1992) 279–283.

[16] H. Minkowski, Diophantische Approximationen, Teubner, Leipzig, 1907. Reprint: Physica-Verlag, Wurzberg (1961).

[17] C. Moore and I. Pak, Ribbon Tile Invariants from the Signed Area, *Journal of Combinatorial Theory, Series A*, **98** (2002) 1–16.

[18] C. Moore, I. Rapaport and E. Remila, Tiling groups for Wang tiles, Proceedings of the 13th annual ACM-SIAM Symposium On Discrete Algorithms (SODA) SIAM eds (2002) 402–411.

[19] I. Pak, Ribbon tile invariants, *Transactions of the American Mathematical Society*, **352(12)** (2000) 5525–5561.

[20] I. Pak, Tile invariants: new horizons, *Theoret. Comput. Sci.*, **303** (2003) 303–331.

[21] O. Perron, Uber luckenlose ausfullung des n-dimensionalen raumes durch kongruente wurfel, *Math. Zeit.*, **46** (1940) 1–26, 161–180.

[22] S. K. Stein and S. Szabo, Algebra and Tiling; homomorphisms in the service of geometry, *Carus Mathematical Monographs*, **25** (1994).

[23] Z. Su and R. Ding, Tilings of orthogonal polygons with similar rectangles or triangles, *J. Appl. Math. & Computing*, **17(1–2)** (2005) 343–350.

[24] S. Szabo, Topics in factorization theory, *Texts and Readings in Mathematics*, **29**, Hindustan Publishing Co. (2004).

[25] W. P. Thurston, Conway's tiling groups, *American Mathematical Monthly*, **97** (1990) 757–773.

[26] S. Wagon, Fourteen proofs of a result about tiling a rectangle, *Amer. Math. Monthly*, **94** (1987) 601–617.

Symmetry: A Multi-Disciplinary Perspective No. 16, 2011, pp. 119–122.

Dirac String Trick: A Demonstration to Reveal the Topology of $SO(3)$

Amol Deshmukh

IISER, Mohali
e-mail: ms07004@iisermohali.ac.in

Abstract. This brief article describes a model to unearth and demonstrate the topological properties of the group $SO(3)$. The fact that 2π rotations are topologically distinct from no rotation, whereas the rotation by 4π is equivalent to no rotation, has important consequences for physics. This demonstration is aimed at bringing this fact out for a general audience in an explicit and visual manner.

Keywords. Homotopy, spin-1/2 particles, rotation group, Dirac's string.

The topological properties of groups have fascinating features and can have non-trivial, counter-intuitive consequences for physical systems. The group $SO(3)$, the group of rotations in three dimensions, plays a very important role in physics. The particular fact that a rotation by an angle 2π is topologically distinct from "no rotation" and it is a rotation by 4π which in fact "really brings things back" has implications for properties of quantum systems. We describe here a demonstration based on two cubes with their corners connected by strings, to bring out this particular topological property of $SO(3)$. Via this demonstration, one can see how 4π rotations are reducible to no rotation by continuous manipulations, while the 2π rotation is not reducible to "no rotation" by such manipulations. The demonstration is designed to be used as a teaching aid to bring out this fascinating property of the group $SO(3)$.

The group of all proper rotations in 3-dimensional space, which is also known as $SO(3)$, is the most useful non-abelian continuous group used in physics. The group $SO(3)$ is defined as the group of all continuous linear transformations in 3-dimensional Euclidean space which leave the length invariant. The group $SO(3)$ stands for the set of all 3×3 special orthogonal matrices. The topological properties of $SO(3)$ group are very different from the familiar topologies. We make use of this topological property to show that an object when connected to its surroundings by a belt or strings gets entangled with those connections after its rotation by '2π', but amazingly those connections get disentangled after a '4π' rotation.

In nature, there are a few elementary particles with spin $\frac{1}{2}$ (electron for example), for which the state after 360° rotation is different from the unrotated state. We would need a 720° rotation to get back to the same (i.e. unrotated) state for these particles. Obviously, the $SO(3)$ group is incorporated in the physical theories to describe this behavior of spin particles. This phenomenon was often cited by Paul Dirac to explain

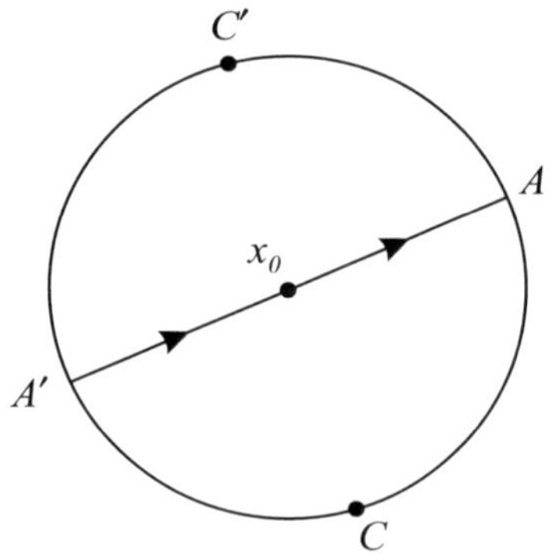

Figure 1. Section of a solid sphere, which represents the topological space corresponding to the group $SO(3)$.

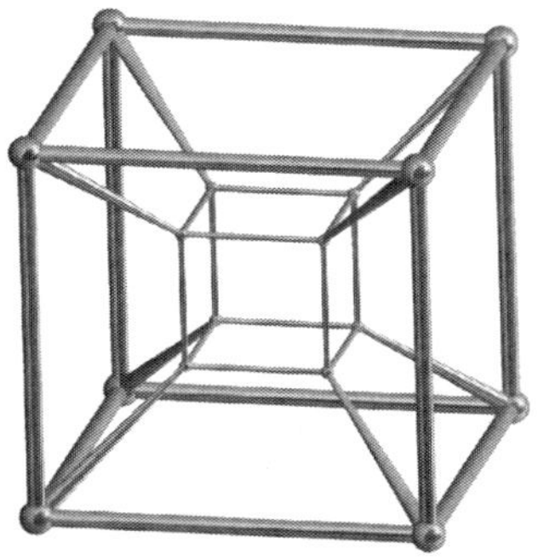

Figure 2. Picture of a physical model with a cube within a cube and the corners of the cubes joined by strings. This model can be used to demonstrate the topological difference between the 2π and 4π rotations.

why a rotating body can have angular momentum of half a quantum, but of no other fraction.

To construct a topological model of $SO(3)$, consider Cartesian co-ordinate system with a totally unsymmetrical body pivoted at the origin. The unrotated state of the body can be marked against the fixed co-ordinate system. Now, there is a total freedom for the body to orient itself in any direction, only constrained to the condition that it's origin is fixed. There is a well known theorem in Mechanics, known as Euler's theorem, which says that "the general displacement of a rigid body with one point fixed is a rotation about some axis". It means that, any rotated state can be specified using an axis and the amount of rotation about that axis. If we consider any arbitrary axis in 3-dimensional space, about this axis any rotation from 0° to 360° is allowed, which will correspond to different rotated states of the object. For the topological model we are going to construct, consider a solid sphere of radius 'π' in an abstract space. With this solid sphere, any rotation in physical space can be specified by a vector pointing along the axis of rotation, with a length equal to the amount of rotation in clockwise direction. So, it is obvious that the two points corresponding to a rotation by 'π' about any axis and rotation by 'π' about the opposite axis, are same. Thus,

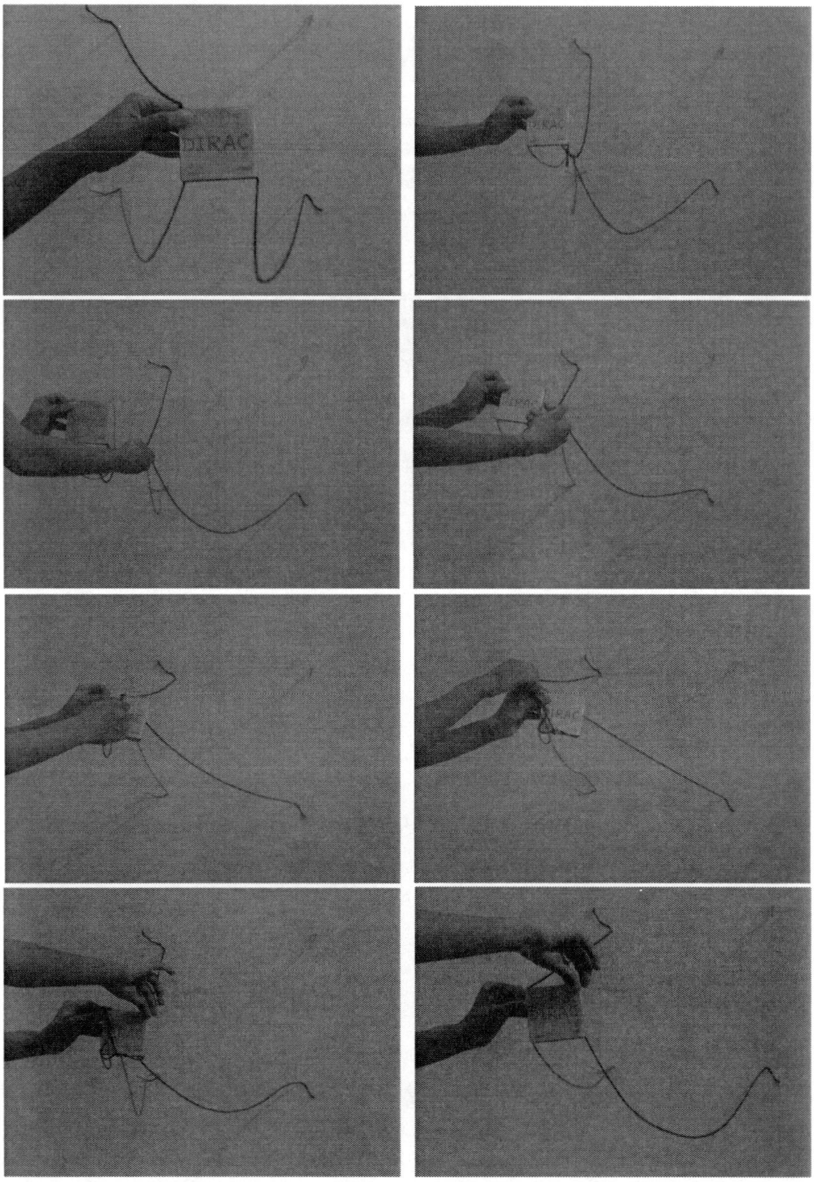

Figure 3. The steps involved in disentangling the cubes after a rotation by 4π. The top left picture shows the unrotated placard and the top right picture shows the placard rotated by an angle 4π (left top). The subsequent pictures show the steps needed to unentangle the strings and in the last picture (bottom left) the placard is back to the original state of "no rotation".

to complete the topological model we can identify (or glue together) these antipodal points in the abstract space.

Figure 1 is a section of a solid sphere described above. The antipodal points such as A and A' (or C and C') are identified, or in other words, they represent exactly the same rotation. As we start rotating any object in real space, a smooth path which starts from the origin is generated in the topological model as shown in the Fig. 1. Because the points A and A' are the same points, after reaching point A, the path is traced from A' till origin. Thus, the path AA' corresponds to the 2π rotation of object in real space. Interestingly, this path can not be smoothly deformed to a point but if we re-traverse path AA' again, the total path which corresponds to 4π rotation of an object in real space, can be smoothly deformed to a point or mathematically, this path is homotopic to the path corresponding to no rotation.

A physical model can be constructed to display this phenomenon of $SO(3)$ group. Take two cubes with different sizes and tie every corner of the smaller cube to the corresponding corner of the larger cube, as shown in Fig. 2.

The demonstration can described via the following steps:

- **Step 1:** Let us call the state in which every corner of small cube is tied with the corresponding corner of the bigger cube as an initial state.
- **Step 2:** Hold the inner cube in your hand as shown in Fig. 2 and rotate this inner cube by 360° about an axis perpendicular to the plane of the paper. The strings get entangled. If we now try to disentangle the strings without rotating the inner cube we will not succeed (i.e. we will not be able to bring it back to the initial state.).
- **Step 3:** But if we now rotate the inner cube in step 2 by another 360°, so that the total rotation becomes 720°, the strings can now be disentangled, without rotating the inner cube. The process by which we can disentangle the cubes is described in Fig. 3.

A series of pictures is given in Fig. 3 which shows the steps involved in the disentanglement of the cords. For convenience, only four strings are shown in pictures. But the steps involved in the model with eight strings are same.

Acknowledgments

I would like to give my special thanks to Prof. Arvind for introducing me to this topic, for guidance and valuable discussions. I would also like to express my gratitude to Prof. Sathyamurthy (Director, IISER-Mohali), Prof. I.B.S. Passi (Convener) and the organizing committee of the symposium, for giving me an opportunity to present the poster in the symposium.

References

[1] Ethan D. Bolker, The Spinor Spanner, *The American Mathematical Monthly*, **80(9)** (Nov., 1973) 977–984.

[2] Group theory and physics, S. Sternberg, Cambridge University Press (1994).

Symmetry: A Multi-Disciplinary Perspective No. 16, 2011, pp. 123–129.

Symmetry and Probability: The Principle of Indifference

Rajni Ranjan

IISER Mohali, Sector-81, Knowledge City, SAS Nagar, Mohali 140 306, Punjab
e-mail: ms07018@iisermohali.ac.in

Abstract. The principle of equipossibility is the basic principle underlying the connection between symmetry and probability. This article emphasizes the thought that symmetry of an object depends on the way it is viewed. In absence of an explicit mention of the frame of reference, the questions on symmetry may lead to various conflicting paths. As supporting examples, problems from probability are considered. Thus it is as important to understand the meaning of symmetry in a given context as to understand the meaning of random, say in probability. Measure theory acts as a connecting link between the two notions.

Keywords. Principle of indifference, symmetry and measure, Bertrand's paradox.

Introduction

Symmetry refers to the property of balance and well defined self similarity and can be observed with respect to space or time. Mathematically, this notion is defined by the language of groups and their actions. In another words, if an object appears the same on applying some transformation, then the transformation is said to be a symmetry of the object. A set is used to describe the elements of the object and the symmetry group of the object is the set of bijections of this set. As an example, let us consider a 2 dimensional equilateral triangle. There are exactly six different transformations, including the identity transformation which would give us the same figure. On computation, its symmetry group, which is of order 6 comes out to be the symmetric group on 3 elements i.e. S_3.

A group G is said to *act* on a set S if there exists a map $\phi : G \times S \to S$ satisfying the following properties:

1. $\phi(1, s) = s$
2. $\phi(g_1 g_2, s) = \phi(g_1, \phi(g_2, s))$.

The language of group actions is powerful in the sense that it helps us to apply the geometrical ideas to abstract objects.

Probability is a way of expressing knowledge or belief that an event will occur in future. There is a strong connection between symmetry and probability in the real world. Often, we rely on symmetry to infer the possibilities of various outcomes. In doing so, the classic principle of indifference is used, which states:

> If there are n possible outcomes and there is no reason to believe that any one of them is more likely to be true than the others, then they are assigned equal probabilities.

This principle has several manifestations in various situations that have to be considered while calculating the probabilities. The exploitation of this principle traces back to Laplace. In 1812, he published his famous book *Théorie analytique des probabilités* in which he laid down many fundamental results in statistics. He had also previously indicated the method of estimating the ratio of favourable cases to the whole number of possible cases by the help of generating functions in his 1779 paper. Based on the principle of indifference, he derived the 'rule of succession'. It states that if an event has only two outcomes i.e. success and failure, then under the assumption that little or nothing is known about the relative plausibilities of the outcomes, the conditional probability that the next trial is a success, given all the previous trials were successes is given by

$$P(\text{next trial is a success}) = \frac{s+1}{n+2}$$

where s is the number of previously occurred successes and n is the total number of trials observed.

Using this rule, he derived that the probability that the sun will rise tomorrow; given that it has risen everyday for the past 5000 years is approximately 1826250/1826251.

Classical probability theory

A modified version of the principle of indifference involves the following principles:

1. **Principle of Uniform Distribution** which is to assume a uniform probability distribution in the absence of reasons to the contrary.
2. **Principle of Stochastic Independence** which is to assume statistical independence of the events in absence of reasons to the contrary.

To emphasize the significance of the principle of indifference, let us consider the classical example of a die. Dice are usually cubes of a homogeneous material. Symmetry suggests that a 'fair' die has the same chance of landing on each of its six faces when tossed. In fact, the four other regular solids, the tetrahedron, octahedron, dodecahedron and icosahedron satisfy this criterion and are hence also fair. It means that any face can be transformed into any other by a rotation. Mathematically, it is the same as saying that the rotation group of the polyhedron acts transitively on the faces. Hence, it is justified that each face is assigned the same probability of showing up when such a die is rolled. In fact, in experiments, it has been observed that each face lands approximately the same fraction of the times. The probability of any event in the classical approach is the ratio of the number of favorable to the number of total outcomes. So the probability that the number 3 will occur on a single roll is estimated to be 1/6.

In the above example, judging that the situation under consideration appeals to the principle of indifference is fairly easy, but it is not the case in general. Although the principle applies in such situations too, categorising the 'symmetrical' situations is in general, not trivial. In fact, the symmetry of the situation might just be apparent. For example, in the case of the dice, they might be loaded and still appear perfectly symmetrical. It needs a careful investigation of the case as to which parameters are to be considered and which are not in assigning probabilities. In fact, in the absence of information about the state of the die, the inferences are most likely to be different for different observers.

At this point, it would be useful to remark that the underlying 'measure' used in calculating these probabilities is the counting measure. It works fine in cases with finite sample spaces, but as we shall see later, this might not always be the case, i.e. the sample space might be infinite or uncountable.

Abnormalities

Many a times, while assigning objective probabilities to certain cases, the principle of symmetry appears to fail. As an example, let us consider a circle and an equilateral triangle inscribed on it. What is the probability that a chord randomly selected on the circle will be longer than the side of the equilateral triangle inscribed on it?

Three of the possible ways of assigning random chords (the dotted lines representing the favourable cases) are:

1. Considering uniform distribution on the perpendicular distance of the chord from the center of the circle and the mid-point of the chord. In this case, the probability is 1/2 as half of the cases are favourable.

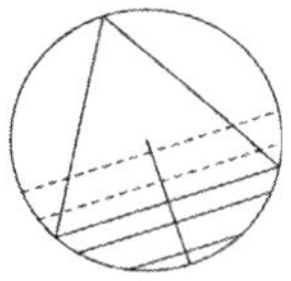

2. Considering uniform distribution on the distance from a fixed point along the circumference. The probability in this case is 1/3.

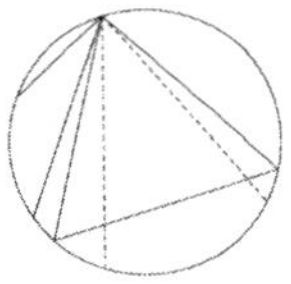

3. Considering uniform distribution on the center of the chord distributed over the interior area of the circle. The probability in this case is 1/4.

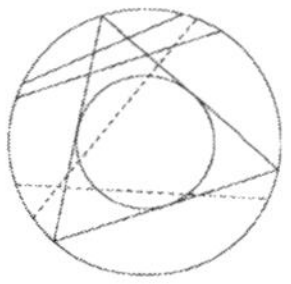

Now, we have three different numerical values of probabilities for the same event. This apparently paradoxical situation is known as the *Bertrand's Paradox*. The paradox is *apparent* because although the sample space is same in all three cases, the measure may be chosen to be any three of them as the question does not specify 'randomness' in any particular aspect.

As another example, let us consider the one of de Finetti lottery. Bruno de Finetti, an Italian probabilist and statistician argued that we should be able to make sense of a lottery in which there are countably infinite tickets and the probability of winning of each of the tickets is the same. The argument seems plausible enough but the difficulty is that there is no way to assign a finite positive value to each number such that the countable sum is 1. It is possible if the higher ticket numbers are assigned a lower probability, but the symmetry considerations make such a lottery unreasonable. One of the possible approaches would be to use relative probabilities instead of absolute ones. In such a situation, it would be reasonable to say that one ticket is 'as likely' to win as any other, without in effect assigning a numerical values to the proababilities.

Now, we consider a cube factory which manufactures cubes of length varying till 2 cm. What is the probability that a cube has length ≤ 1 cm given that it was manufactured by that factory? [F, p. 303]

Three of the possible approaches which one may consider to solve the above problem is to consider uniform distribution in length, area of each face and the volume of the cube. In the first case, the probability is 1/2, in the second 1/4 and in the third, 1/8. One can explain this apparently paradoxical situation by adopting another approach that is the measure theoretic approach of probability.

Measure theory: An introduction

One of the most fundamental notions of probability theory is that of a probability measure. To define the same, we first need to define a σ-algebra.

Given a non-empty set X, a non-empty collection, Σ, of subsets of X is called a *σ-algebra* if

1. $A^c \in \Sigma$ for all $A \in \Sigma$.
2. For a countable family $(A_i)_i \in \mathbb{N}$, $A_i \in \Sigma$, we have $\cup_i A_i \in \Sigma$.

A *finitely additive probability measure* on a σ-algebra is a non-negative real valued function P defined on Σ such that $P(X) = 1$, $P(\Phi) = 0$, and if $A, B \in \Sigma$ are disjoint, then

$$P(A \cup B) = P(A) + P(B).$$

The above measure is said to be *countably additive* if for any sequence $A_i, i \in \mathbb{N}$ of pairwise disjoint sets,

$$P(\cup_i A_i) = \sum_i P(A_i).$$

Any element of Σ is called a *measurable set* in X. The pair (X, Σ) is known as a *measure space*.

A σ algebra corresponds to the events where probability can be "determined". For example, in rolling of a die, it is natural to consider the set $X = \{1, 2, \ldots, 6\}$ as the outcome of the experiment is one of theses six numbers. However, we could bet beforehand that either 1 or 2 will appear. The set of all such statements about the outcome which can be made with certainty is a σ-algebra. In fact, in this example, the σ-algebra is the set of all subsets of X. If we are concerned only about the even and the odd outcomes, the σ-algebra $\Sigma = \{\{1, 2, 3, 4, 5, 6\}, \{1, 3, 5\}, \{2, 4, 6\}, \Phi\}$ would also suffice.

We define a *probability space* as a triple (X, Σ, P) where X is a set, Σ a σ-algebra defined on X and P the probability measure. The σ-algebra corresponds to the events i.e. the collection of measurable sets in the sample space. In this approach, the problems of calculating discrete probabilities corresponds to taking the counting measure on spaces.

Measure theory to the rescue

Using the above approach, several cases which were earlier difficult to assign probabilities to, can be resolved. In cases where the 'number' or 'fraction' of favourable cases is impossible to find, one may consider the measure of the set in the σ-algebra which corresponds to the event under consideration. At this point, it is important to emphasize that probability theory in itself does not provide a natural way of solving the real life problems. Given a σ-algebra of events, there is no canonical way to assign a probability measure on it. Moreover, there is no proper reason to prefer one probability measure over the other. Therefore, for a real life problem, the task of assigning a probability is ad-hoc. However, the probability measure which does not assume any extra information about the experiment should be preferred. This is what we call the principle of "maximum ignorance". It is this principle which forms the basis of the principle of indifference. This principle of indifference has a connection to the action of a particular group on the σ-algebra. For example, let us consider the σ-algebra generated by all open sets of the real line. The *Lebesgue measure* of any interval (a, b) is defined as

$$\mu(a, b) = b - a.$$

This measure is translation invariant as

$$\mu((a, b) + k) = \mu(a + k, b + k) = b - a = \mu(a, b).$$

The translation $(a, b) \mapsto (a, b) + k$ can be viewed as a group-action of real numbers on the above σ-algebra. Thus, identifying the group which acts on the σ-algebra and studying the properties of such group actions can provide insight about the measure which would be most suitable for the problem under consideration.

In the cube-factory example, one of the solutions proposed by Van Fraassen [F, p. 309] is based on the above principle. In this case, the quantities under consideration are not closed under addition and subtraction, which is required for translation invariant measure to be used. Also, the solution to the problem should be independent of the unit i.e. the scale of measurement used. The three equivalent renditions of the problem statement require the solution to be invariant under dilations. Clearly, the Lebesgue measure is not scale invariant. In fact, it has been illustrated using probability density arguments that the log uniform measure is the unique scale invariant measure [M]. Thus, Van Fraassen argues that *log uniform* measure on the real numbers be used to calculate the probabilities. It is defined as:

$$\mu(a, b) = \log b - \log a$$

where $a, b \in \mathbb{R}$. The above measure is scale invariant as

$$\mu(ka, kb) = \log kb - \log ka = \log b - \log a = \mu(a, b).$$

It is important to note that the above measure can only be used for positive quantities as logarithms of negative values are not defined.

On applying this to our problem, we obtain

$$P(1 \leq L \leq 2) = \mu([1,2])/\mu([1,4]) = \log 2/\log 4 = \log /2 \log 2 = 0.5$$

$$P(1 \leq A \leq 4) = \mu([1,4])/\mu([1,16]) = \log 4/\log 16 = 2\log 2/4\log 2 = 0.5$$

$$P(1 \leq V \leq 8) = \mu([1,8])/\mu([1,64]) = \log 8/\log 64 = 3\log 2/6\log 2 = 0.5$$

where L, A, V denote length, area of a face and volume of the cube. We get equal probabilities for all the three interpretations, i.e. 0.5. Hence, the answers are now in agreement with each other. Again, we do not claim the value 0.5 to match the value obtained by actually performing the experiment. There is no evidence that the scale invariant measure we have considered does model the actual experimental set-up.

Now, we move again to the Bertrand's paradox. There are many interpretations to this problem. For instance, in his 1973 paper The Well-Posed Problem [J], Edwin Jaynes proposed a solution to Bertrand's paradox, based on the principle of indifference. Jaynes pointed out that Bertrand's problem does not specify the position or size of the circle, and argued that therefore any objective solution, if it exists, must be "indifferent" to size and position. In other words, the solution must be both scale invariant and translation invariant.

We assume the density function in the interior of the circle to be $f(r, \theta)$, where r and θ are the polar coordinates of the center of the chord. The rotational symmetry in the problem implies that this density is independent of θ.

Now, the function $f(r)$ needs to determined according to the normalization requirement:

$$\int_0^{2\pi} \int_0^R f(r)\, r\, dr\, d\theta = 1.$$

Now, using the scale invariance property, the general solution is found to be

$$f(r) = \frac{q r^{q-2}}{2\pi R^q}$$

where q is a real, positive number which cannot be determined by the scale invariance property. The above observation eliminates the second method described in the question as in that case, the distribution function would be of the form $(R^2 - r^2)^{-\frac{1}{2}}$.

Now, applying the translational invariance, the value of q is determined to be 1. The third method is also eliminated as if straws are randomly tossed onto one circle, then the distribution might not necessarily have the same properties for a slightly displaced circle. Thus, he inferred that out of the three, only the first method is both scale and translation invariant and obtained the density function

$$f(r, \theta) = \frac{1}{2\pi R r}$$

where r and θ are defined as above.

Inference

Some questions in probability may seem paradoxical if they are not properly framed. Different ways of describing the outcome space may yield different probabilities for the same event. Hence while studying such questions, along with the sample space, an adequate knowledge of the measure is therefore essential to avoid ambiguities. It is the careful investigation of the underlying symmetry of the case under consideration which results in a 'good' success rate in the inferences in real-world problems. In lack of experience of outcomes under various circumstances, we have no reason to prefer any distribution over others.

Hence, despite the ambiguities, the influence of symmetry in probability theory as the principle of indifference still persists.

Acknowledgements

I am highly grateful to Dr. Amit Kulshrestha for his invaluable guidance and support throughout. His suggestions helped me improve the quality of the work substantially. I am also thankful to Prof. N. Sathyamurthy, Prof. I.B.S. Passi and the entire organising committee of the Symposium on Symmetry, Feb. 19–20, 2010 for giving me an opportunity to present the poster in the same.

References

[F] B. Fraassen, Laws and Symmetry, Oxford University Press, New York, USA (1989).

[S] S. Stigler, The History of Statistics, Harvard University Press (1986).

[J] E. Jaynes, The Well Posed Problem, *Foundations of Physics*, **3** (1973) 477–493.

[P] S. Portnoy, A Lewis Carroll Pillow Problem: Probability of an Obtuse Triangle, *Statistical Science*, **9** (1994) 273–284.

[N] J. North, Symmetry and Probability, *Philosophy of Science Association: Contributed papers*, 2006. Available online at: http://philsci-archive.pitt.edu/2978/1/symmetry_psa.pdf

[M] P. Milne, A Note on Scale Invariance, *British Journal for the Philosophy of Science*, **34(1)** (1983) 49–55.

Made in United States
North Haven, CT
05 November 2026

21778847R10078